Special limited edition.
/2500

NORTH AMERICAN Endangered & Protected SPECIES

A collection of 80 CITES-listed species
from non-threatened to extinct

Volume One
Limited Edition

ISBN: 1-885743-00-9 (Volume One of six-volume set) and 1-885743-01-7 (six-volume set)
Library of Congress Catalog Card Number: 94-077816

Text and art direction by E. A. Lusch.
Editing by Mark Flint.
Design by Carl S. Darnell.
Project Coordination by The Consultants Group, Eagle Point, Oregon, USA.
Printed in Korea through Jinno International, New York.

Limited edition printing March, 1995.

Table of Contents

Acknowledgments

North American Endangered & Protected Species is a labor of love, of caring and commitment. Many people have given long hours to the project without any guarantee that they would receive compensation for their efforts. They gave their time and talent because they believed in the project from the start. They also felt the risk of involvement in a project that may or may not come to fruition was worth the sacrifice. The opportunity to be involved in the creation of an exceptional book of art that could provide ongoing and permanent help in the effort to eliminate illegal taking of endangered species was simply too tempting to refuse.

In short, they had very much to gain, and only time and money to lose.

Now that the first volume of this series is a reality, those people deserve credit for their time, their talent and their tenacity. Special thanks to:

—Dr. Ralph E. Wehinger, who conceived the project, pulled the people together to make it happen, and provided the glue to keep things together even when events conspired to pull the entire project apart. Joining Dr. Wehinger in this effort was his family, particularly his wife, Laura, who put up with his long absences as he worked on the project and offered her support and creativity when problems seemed insurmountable.

—The "core" team that actually put the book together. Eighteen highly talented wildlife artists created stunning portraits of wildlife, and then waited patiently while the rest of the team put the words and art together and attended to the myriad details of book publishing.

Ed Lusch pulled the artwork and text together, Mark Flint edited the text and Carl Darnell designed the book and laid out the text.

The three core team members actually were a team of six. Linda Lusch spent many hours helping her husband with typing and research; Cindy Darnell helped Carl with design as well as running the family business single-handed while he labored over the book; and Chloe Becca was a stalwart proof-reader and candid critic for her husband, Mark Flint, during the editing process.

—The Honorable Izgrev Topkov, CITES Secretary General, for lending his support to the project. That support unlocked many doors which never would have opened.

—The Conservation Treaty Support Fund, which provided the original organizational "umbrella" for the project.

—The United States Department of the Interior, U.S. Fish and Wildlife Service, Division of Law Enforcement, National Fish & Wildlife Forensics Laboratory in Ashland, Oregon, for research assistance.

—Rolf Wallenstrom, retired Regional Director for the U.S. Fish and Wildlife Service, for providing vital liaison with federal agencies during the research process.

—Attorney and project supporter Steven Jamieson, of Frohnmayer, Deatherage, Pratt, Jamieson and Turner, whose excellent counsel, guidance and advice helped iron out numerous bumps in the road before they appeared.

This book, and the resulting funding for wildlife law enforcement, has come to fruition because of a broad base of support. Those we have listed are but a few of the many contributors whose efforts may not have been as large, but are no less important.

We are proud to have worked with you.

Dr. Ralph E. Wehinger
Carl Darnell
Mark Flint
Ed Lusch

Preface

A plant or animal species can be said to exist in one of four population conditions: it is either flourishing, holding its own, threatened, or endangered. The final stage is extinction, a permanent erasure from the ecosystem. The price of extinction is one we can often never assess; we will never know what benefit many vanished species might have provided man.

Most medicines, many of which we call "miracle drugs," are derived from plant or animal species. A new cure for a current disease, or even one that we haven't yet experienced, could be "just around the corner," waiting to be discovered in a plant or animal species.

Can we afford to let any species perish? In essence, that is what CITES is about: maintaining the viability of plant and animal species for their inherent value and for man's immediate and unforeseen uses.

Achieving the CITES mandate requires funds. The CITES identification manuals, for example, are in need of ongoing revision and expansion, and funding for that critical task will be provided from sales of this book. *North American Endangered & Protected Species* is the first of a series of limited edition volumes dedicated to protected wildlife. When completed, the series will cover major protected and endangered species throughout the world. Proceeds from the sale of this series are being contributed to the International Wildlife Recovery Center and the Conservation Treaty Support Fund for the benefit of CITES.

This first volume of wildlife coffee table artwork features 80 illustrations of North American endangered plants and animals. Each illustration has been meticulously rendered by some of the finest wildlife artists of North America.

The focus of these artists was to paint with absolute biological integrity in their own inimitable styles, and they have achieved this with magnificent results.

But this book is more than a showcase of wildlife art. It is a "living trust" in that it will do more than simply increase awareness and appreciation for the plants and animals that inhabit our planet. *North American Endangered & Protected Species* is designed to be an important source of revenue which will ultimately help promote the security of the earth's plant and animal resources.

An Introduction to the Convention on International Trade in Endangered Species of Wild Fauna and Flora (CITES)

Illegal trade in wild plants and animals is big business on an international scale, at least five billions dollars annually. The facts of illegal trade are sadly sobering. The illicit sale of products from these animals and plants is second only to worldwide sale of illegal drugs in dollar volume.

Illicit trade in wild plants and animals constitutes a global threat to the existence of thousands of species. Leather, feathers, medicine, the pet trade — any perceived commercial wildlife value will be exploited, and if left unregulated, exploited to the point of endangerment or extinction.

Legally traded wildlife species can be an important part of a country's economic base, and protecting certain species may depend entirely upon those species' economic importance. It is a fact of life that some wildlife species will be protected only if there is an economic incentive to do so.

The overwhelming need to control the trade in wild plants and animals to ensure their survival prompted an international treaty in 1973 aimed at monitoring, restricting, and in some cases eliminating the trade in internationally traded wildlife species.

The Convention on International Trade in Endangered Species of Wild Fauna and Flora — CITES — became operational in 1975 and today CITES has more than 120 signatory countries.

The CITES objective is "The rational and sustainable conservation of the earth's living resources for their own sake and for human benefit."

To foster that objective CITES acts as the international regulatory body to adjust, monitor, and if need be, prohibit commercial importing and exporting of wild animal and plant species that are, or may become threatened or endangered by trade.

CITES was not intended to ban all trade in wildlife, but to make sure that trade does not threaten the existence of any wild plant or animal.

CITES helps control the international trade in wildlife by requiring government permits for trade in any CITES listed species. Permits for commercial trade are issued only if a species is not listed in Appendix I of CITES, and trade does not pose a threat to its existence as a species. Trade for scientific purposes may be permitted for species listed in Appendix I.

A CITES listed species is dependent upon its listing in one of three appendices:

—Appendix I listing indicates most endangered status. International commercial trade in Appendix I species is outlawed. Permits for trade are granted only for scientific purposes, such as trade between zoological institutions to maintain effective gene pools through captive breeding.

—Appendix II listing means the species is not presently threatened with extinction but could become so with unregulated international trade. Permits for Appendix II species may be issued as long as trade will not adversely affect the species survival. In certain cases some non-threatened "look-alike" species are included in Appendix II to prevent similar looking threatened species from being traded. It is this "look-alike" listing which causes much controversy from within and without the CITES ranks.

—Appendix III shall include all species which any member country identifies as being subject to regulation within its jurisdiction.

To achieve CITES goals, member countries are required to submit annual reports of trade records to the CITES Secretariat. CITES regulations are enforced by customs officers and in some cases fish and wildlife agents. Each member country is responsible for establishing management and scientific authorities to collect data on wildlife imports and exports; that data is submitted to the Secretariat for monitoring. The Secretariat rules on and issues permits, determines whether trade will endanger a species and implements CITES' many provisions and resolutions of its Conference of Parties. The Secretariat also works with the Wildlife Trade Monitoring Unit and TRAFFIC (Trade and Analysis of Fauna and Flora In Commerce) to identify problems in wildlife trade and recommend solutions.

CITES' key responsibility is controlling the international trade in endangered wildlife, but in so doing a broad range of wildlife-friendly activities also occur.

CITES member nations meet every two to three years, and the gathering of the now more than 120 signatory nations becomes a major international conservation forum. The countries update and revise lists of endangered species, and share updated studies and monitoring of endangered species. The meeting of these countries in an international conservation forum provides a great amount of publicity, which adds to public awareness of wildlife issues. Developing countries gain valuable insight and training in wildlife management, and greater understanding of the need to protect their resources.

The rational and sustainable conservation of the earth's living resources for their own sake and for human benefit.

-CITES objective

CITES assesses captive breeding and wildlife ranching programs to relieve pressures on certain wild species, and addresses these programs as conservation tools. Through CITES, endangered species collections among zoos can be coordinated and research information exchanged. The reintroduction of endangered captive bred species into the wild is also aided by CITES coordination efforts.

At the biennial conferences CITES considers proposals by member countries to amend the appendices. Amendments may add a species, delist a species, or change a species' Appendix designation. Decisions to amend the Appendix must be based upon the best scientific data available and agreed to by two-thirds of the parties present and voting. To that end CITES depends upon information provided by the World Conservation Union (IUCN) specialists. The IUCN comprises over 100 such specialist groups. In addition to the IUCN's groups, CITES depends upon the TRAFFIC Network, a joint effort between the World Wide Fund For Nature and IUCN, and the WCMC — The World Conservation Monitoring Center. These major conservation organizations ensure CITES' credibility by providing reliable and current information on the trade and biological status of species in question.

CITES is additionally supported by the Conservation Treaty Support Fund and its affiliates, which promote public awareness of conservation treaties like CITES and raise financial support.

It is important to remember that the control of trade in more than 40,000 CITES listed species is to fulfill the long term objective: "The rational and sustainable conservation of the earth's living resources for their own sake and for human benefit, which must be protected for this generation and the generations to come." CITES uses direct and indirect measures as it seeks to achieve that goal.

About the International Wildlife Recovery Center

The International Wildlife Recovery Center (IWRC) is a non-profit organization originally formed in southern Oregon to create, develop, manage and operate a permanent wildlife rehabilitation facility capable of handling, cleaning, treating and providing medical care for wildlife affected or contaminated by oil and/or hazardous materials, and to educate the public concerning wildlife issues.

Dr. Ralph E. Wehinger, the founder of IWRC, was the driving force behind the effort to establish the Mark O. Hatfield Environmental Science Center at Southern Oregon State College in Ashland, Oregon. Dr. Wehinger led the effort to locate the National Fish and Wildlife Clark R. Bavin Forensics Laboratory in the science complex as its first tenant. Dr. Wehinger also founded the Pacific Northwest Museum of Natural History, also located at the Hatfield Environmental Science Center.

IWRC has expanded its mission to include international fish and wildlife forensics education, and to fund needed updates to identification manuals for wildlife law enforcement and customs officials. The series of books, of which *North American Endangered & Protected Species* is the first volume, is the vehicle for establishing a trust fund that will provide that funding on an ongoing basis.

INTERNATIONAL
WILDLIFE
RECOVERY

Members of the Board of Trustees of IWRC are:

Jaguar

Panthera onca
Appendix I

This magnificent two-hundred-pound-plus cat is shy by nature. That reclusiveness, combined with its nocturnal habits and a preference for dense jungle habitat, makes population estimates extremely difficult throughout its range.

The Jaguar is believed extinct in the southern United States, and rare through Central and South America.

The demise of the Jaguar stems from many sources, all of which continue to pose serious threats to its existence. The destruction of its habitat by agricultural practices, road building, and other developments continues to be this cat's biggest obstacle to survival.

But a host of other problems threaten this species. Jaguar pelts are extremely valuable and, while prohibited from trade by CITES, a black market trade continues.

Jaguar are considered a big-game trophy species in some countries. Wildlife exhibitors and zoos prize them for their beautiful looks. And they are shot by ranchers as threats to cattle.

To ensure the rather tenuous future of Jaguar habitat, national Jaguar parks have been established. More such protected parks are needed.

HABITAT:
The Jaguar was once found in dense cover areas of southern Texas, Arizona and New Mexico, south through Central and South America.

Jaguar Panthera onca **Al Agnew**

Prairie Falcon

Falco mexicanus
Appendix II

Cruising almost at ground level, Prairie Falcons capture prey, such as jackrabbits, squirrels, doves and quail, on the ground or in very low flight, unlike other falcon species, which dive at prey from high elevations.

As their name implies, Prairie Falcons, which belong to a group of raptors known as "desert falcons," are a bird of the open plains and desert grasslands. Current estimates put the number of birds at eight to twelve thousand. While the species is believed to be declining worldwide, its population is thought to be stable in North America.

The Prairie Falcon's population is stable, and its Appendix II listing is a result of its look-alike status with the Peregrine Falcon.

Primary threats to Prairie Falcon include pesticides and other agrichemicals and habitat alteration or destruction. Additionally, shooting, especially of young birds, accounts for a high degree of mortality. They are not commonly used in the falconry trade.

In addition to CITES protection, this "falcon of the deserts" is protected under migrating bird laws of Canada, the United States and Mexico.

HABITAT:
Prairie Falcons breed from southeastern British Columbia south to Baja Mexico, the southwestern United States and Texas. Their winter range extends from southern British Columbia and Alberta to Mexico. Prairie Falcons have been sighted in other states and provinces as well.

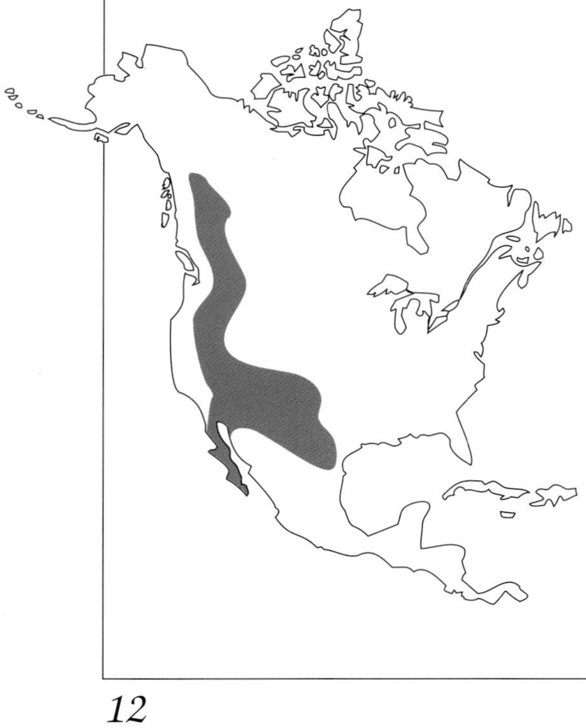

Prairie Falcon *Falco mexicanus* Al Agnew

Gray Wolf

Canus lupus
Appendix I

The Gray Wolf, also known as the Timber Wolf, once ranged over most of North America in close association with its principle prey species: deer, elk, caribou, moose and antelope. During the 19th and 20th centuries, wolves were trapped, shot, and poisoned by the thousands. Government-sponsored poisoning programs and bounties eliminated wolves almost entirely from coast to coast in the United States.

Canadian and Alaskan Gray Wolf populations are estimated to number somewhere around twenty-five thousand.

The total United States population outside of Alaska is probably under fourteen hundred, twelve hundred of them in northeastern Minnesota.

Wolves are protected by Canadian laws, the United States Endangered Species Act (United States populations only), Alaskan and other state fish and wildlife regulations, and CITES. The Gray Wolf is one of more than twenty subspecies of North American wolves.

If wolves are to be allowed modest expansion into certain United States wilderness and national park areas such as Yellowstone and Glacier parks, or the Selkirk Mountains of Idaho, the public must be educated as to the role they play in a healthy wilderness environment. The fear of wolf attack brought about by centuries of myth, fairy tales, and exaggeration must be replaced by insights concerning predator/prey relationships.

The haunting melody of wolf howl once again echoing across the United States wilderness is an exciting prospect. With care and sensitivity, this can be achieved in appropriate wilderness areas.

HABITAT:

United States populations of the Gray Wolf outside of Alaska exist only in northeastern Minnesota, small sections of Michigan and Wisconsin, Montana, possibly Idaho, Washington and Oregon.

Gray Wolf *Canis lupus* **Al Agnew**

Eastern Cougar

Felis concolor cougar
Appendix I

The Eastern Cougars, also known as Puma, Mountain Lion, or Catamount, fell victim to predator control programs, hunting, and habitat destruction; ignorance of the animal's role in nature played a part in all of these factors.

This species is rare and possibly extinct throughout its former United States range, and is rare in its range in eastern Canada. Eastern Cougars are protected under the Endangered Species Act, state and provincial laws and CITES. No international legal trade exists for Eastern Cougar and potential trade threats appear minimal.

Eastern Cougar protective reserves may be important elements in maintaining the possibility of the species survival, but the vast territory required for a cougar makes it difficult to provide adequate preserves. Its look-alike cousin, the Florida Puma, is also a highly endangered species.

HABITAT:
The Eastern Cougar once ranged from New England to the Great Plains as far south as Georgia and Alabama. In Canada it was found as far east as New Brunswick, and as far west as the plains. It is now believed extinct throughout its former United States range, and rare throughout its Canadian range.

Eastern Cougar *Felis concolor cougar* Al Agnew

Snowy Owl

Nyctea scandiaca
Appendix III

A denizen of the far north, the Snowy Owl is heavily dependent upon lemmings as its primary food source. When lemming are scarce, the Snowy Owl may not nest at all. However, if lemming populations are peaking, as many as twelve owlets may be raised by a single pair.

In addition to its customary prey of lemmings and snowshoe hares, the Snowy Owl also will feast on shorebirds, rodents, crows, fish, and carrion. The bird prefers to hunt in daylight — an unusual practice for owls.

When Snowy Owls are forced to migrate south during prey shortages, their weakened state reduces chances for survival. In addition, some are killed by thoughtless shooters and others perish from collisions with automobiles. Still others may succumb from ingesting pelagic birds soaked by oil spills.

The Snowy Owl is — as are all owls — protected by United States and Canadian wildlife laws, and CITES Appendix III.

Despite the rather harsh demands nature places upon the Snowy Owl, the species is holding its own. There is no known trade in the species.

HABITAT:
Ranging farther to the north than any owl species, the Snowy Owl generally will spend the summer as far north as the Arctic Circle, wintering in southern British Columbia or Washington. Snowy Owls will migrate into Washington and Oregon when lemming and snowshoe hare populations have crashed in the far north. During this migration, the Snowy Owl prefers coastal locations, but may at times venture inland. Snowy Owls show up in the eastern United States as well.

Snowy Owl *Nyctea scandiaca* Marian Anderson

American Bobcat

Felis rufus (formerly Lynx rufus)
Appendix III

This feline is highly adaptable to a variety of habitats — scrub, swamps, lowland, highland forests, and deserts — just about every conceivable area in which prey may be found.

The Bobcat weighs in at twenty to thirty-five pounds, is twenty-five to thirty inches long, and features a short, stubby five-inch-long tail. Its fur may vary from reddish brown to a light tawny color with streaks and black spots. It is similar in appearance to the lynx, but can be differentiated by smaller ear tufts, smaller-looking feet, and a white-tipped tail.

The Bobcat is thriving throughout most of its current range, and is thought to be increasing in population, but the cat is rarely seen because of its extremely shy, reclusive nature and solitary nocturnal habits.

The Bobcat's success is due not only to its wide tolerance of habitat types, but also its non-discriminating palate. The Bobcat will eat anything it can catch, from rodents and reptiles to birds and rabbits — even carrion. Generally hunting within a two-mile radius, the cat will vary its range according to prey availability.

Bobcats are still trapped under heavy restrictions and quotas, but because the American Bobcat's fur is impossible to differentiate from the Mexican Bobcat, officials have serious problems enforcing quotas and protecting the more threatened Mexican subspecies. Their look-alike status prompted CITES to list all bobcat subspecies.

HABITAT:
The American Bobcat is the most widespread North American wild cat, occurring throughout the western, eastern, and southern United States, southern Canada, and Mexico.

American Bobcat *Felis rufus* Marian Anderson

Canadian Lynx

Felis lynx
Appendix II

Because of its thick, luxuriant, water-resistant coat, the Canadian Lynx has long been a favored target among fur trappers.

The Lynx weighs up to forty pounds and is characterized by a thick and relatively long light brown or grayish buff fur with darker mottling. It is distinguished from its look-alike cousin, the American Bobcat, by larger ear tufts or points, a thicker cheek mane and a black-tipped tail.

The Lynx's health is closely tied to the snowshoe hare, its primary food source. Any loss of hare habitat or other circumstance affecting hare populations has a direct effect on Lynx numbers. About every ten years the hare population crashes, causing Lynx to migrate into more southerly locales. A collapse in snowshoe hare numbers is followed by a collapse in Lynx numbers.

The Lynx had been trapped to near extinction, but tighter restrictions and the decline in the fur coat markets have aided in its recovery.

Lynx are protected under various Canadian wildlife regulations, United States fish and game laws, and CITES. Of critical concern to the health of the Lynx is protection of their northern forests.

HABITAT:
The species is found throughout Canada, the Yukon Territory, and Alaska. It may also wander south into the Rocky Mountain States if snowshoe hare populations — its principle prey species — collapse in the North.

Canadian Lynx *Felis lynx* Marian Anderson

Saw-Whet Owl

Aegolius acadicus
Appendix III

At six and one-half to eight inches tall, the diminutive Saw-Whet is one of the smallest of North America's eighteen species of owls.

Saw-Whets are tolerant of human activity and will nest close to homes if nest boxes or suitable trees are present.

Possibly the biggest threat to this owl is loss of old snags for nest sites. Remaining sites are often taken over by starlings or screech owls, leaving few tree-type holes or cavities left for Saw-Whets to nest in. Protection of snags is critical to the wellbeing of this species.

The brown, blotched coloring of a typical adult Saw-Whet is generally quite inconspicuous and provides perfect camouflage in the owl's natural habitat. It is easy to stroll by a Saw-Whet and never know it is there.

The species is protected by United States fish and wildlife laws and monitored by CITES. No known commercial trade exists, but they are on occasion illegally taken from the wild for pets. There is no significant threat to the species, and its listing is a part of the global listing of raptors.

HABITAT:
Found throughout the United States and into Canada and Mexico, the Saw-Whet Owl's favored habitat includes mixed growth forests with old snags containing cavities for nest sites.

Saw-Whet Owl *Aegolius acadicus* Joey Berlin

Ocelot

Felis pardalis
Appendix I

This small, spotted, reclusive cat still can be found in Texas and Arizona, but its numbers are dramatically reduced. Ocelot population status is unknown but believed to be declining. The Ocelot's pelt is a high demand item for the fur trade — one coat may bring as much as forty thousand dollars.

Anyone considering wearing an Ocelot coat should first consider that their action is encouragement to poachers, and hastens the species decline.

Ocelot are protected by the Endangered Species Act and CITES, but illegal trade in Ocelot pelts remains high. The illegal use of fur for coats and other products may be directly responsible for much of the depletion of this species.

The Ocelot prefers secluded areas of dense thickets, brush, and desert scrub environments in association with streams. This type of habitat is rapidly diminishing and remaining tracts need to be preserved if Ocelots are to survive in the wild.

Transplanting of Ocelots from less desirable habitat into secluded wildlife refuges also plays a role in this enhancement.

HABITAT:
Historically, the Ocelot lived throughout much of Texas, southeastern Arizona, and New Mexico.

Ocelot *Felis pardalis* Joey Berlin

Margay

Feliswiedi
Appendix I

This little known and extremely rare small cat resembles the ocelot, with which it is often confused.

Beautifully colored yellow with black spots, the Margay is very shy and secretive, making it impossible to determine how many remain.

Margay have been hunted by the fur trade extensively, but due to their resemblance to their cousin, the ocelot, it is difficult to determine how many pelts actually enter the trade market.

While protected from trade by CITES Appendix I listing and the Endangered Species Act, Margay numbers still are thought to be declining in the wild because of illegal hunting for the fur trade.

Deforestation of Margay habitat also poses a threat to the species.

The Margay may no longer exist in southern Texas and the remaining populations in Mexico, Central and South America are severely depleted. Ending the demand for wild cat fur jackets would be the best protection for Margay populations, if available habitat remained. Reintroduction into suitable habitat sanctioned for protection is a possibility.

HABITAT:
The Margay's original range included tropical and subtropical forests in southern Texas through Mexico into Central and South America. Today they may not be found north of the Mexican border.

Margay *Felis wiedi* Joey Berlin

Cooper's Hawk

Accipiter gentilis
Appendix III

A determined predator that will often chase prey on foot into thick brush, the Cooper's Hawk prefers to catch and eat small birds, but also hunts rodents, small mammals, and occasionally lizards and snakes. The hawk may also take barnyard fowl and domestic pigeons from time to time, and can be quite aggressive about it.

The Cooper's Hawk is medium-sized, usually thirty-nine to forty-five inches in length, with a wingspan ranging from twenty to thirty-four inches. As in most hawk species, the female is larger than the male, weighing up to two pounds to the male's fourteen ounces.

Cooper's Hawks prefer open woodlands in lowland elevation where they are often seen perched in tree limbs or fence posts on the look-out for a meal.

A very powerful hawk, the Cooper's Hawk is often mistaken for the sharp-shinned hawk, but can be differentiated from this hawk species by its larger looking head and a white-tipped tail band.

The Cooper's Hawk is reasonably adaptive to agricultural takeover of native habitats and seems to be doing well over much of its range. As with nearly all raptors, pesticides have been a problem in the past, and could pose a serious threat if reintroduced.

Cooper's Hawks are regulated by international laws, state laws, and by CITES Appendix III protection.

HABITAT:
The Cooper's Hawk is found over most of the United States and into southern Canada. They do migrate, but primarily over winter in the United States.

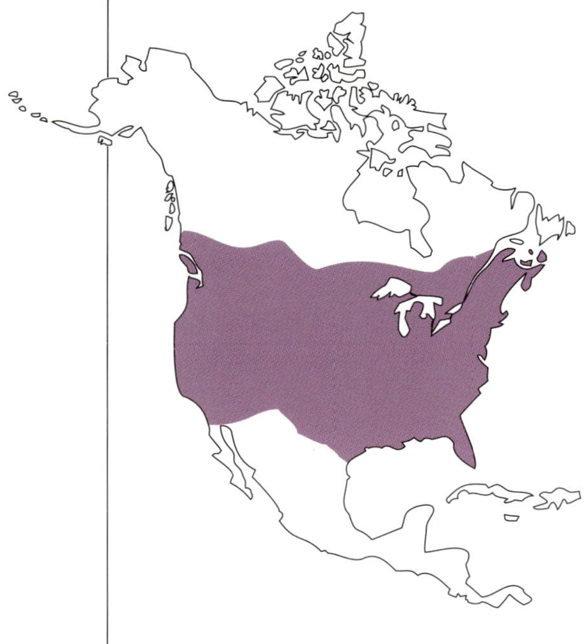

Cooper's Hawk *Accipiter gentilis* Guy Crittenden

Southern Sea Otter

Enhydra lutris nereis
Appendix I

Better known as the California Sea Otter, this marine mammal is completely adapted to its inshore ocean existence. The Sea Otter is able to dive down to two hundred fifty feet while holding its breath for five minutes. Weighing up to eighty pounds, Sea Otters live primarily among the California kelp beds, where they lounge on their backs, breaking open and feeding upon a variety of shellfish species.

Sea Otter were common from the Alaska coast south to Baja, but fur hunters all but exterminated the Pacific Coast populations outside of Alaska. It was thought that only fifty otters remained by 1911, but the discovery of a colony of otters in 1938 expanded that number to about seventeen hundred. Sea Otters have been transplanted into other California coastal waters and the Washington coast. In Alaska the Sea Otter population is healthy.

A continuing controversy exists between commercial fishermen and Sea Otters. Both target the same prey species, and commercial fishermen argue that increasing numbers of otters threaten their livelihood. Indeed, Sea Otters have fairly rapacious appetites, requiring up to fifteen pounds of shellfish a day to sustain life.

Resentment runs high, especially among abalone divers, who claim that Sea Otters have wiped out certain abalone beds.

Sea Otters are protected by a host of laws and treaties, including the United States Endangered Species Act, the Marine Mammal Protection Act, state laws, and CITES Appendix I listing.

However, all the laws and treaties aimed at Sea Otter recovery will be to no avail should a major oil spill occur in otter habitat. There is a tremendous amount of oil tanker traffic through the otter's territory, and a large spill could wipe out the otter populations caught in the muck. The U.S. Fish and Wildlife Service is now attempting to stock otters in island areas away from possible oil spills.

HABITAT:
The Sea Otter's habitat includes the Alaska coast through Washington, California and Baja.

Southern Sea Otter *Enhydra lutris nereis* Guy Crittenden

Gopher Tortoise

Gopherus polyphemus
Appendix I

The Gopher Tortoise spends much of its time underground, and often digs into and expands gopher burrows to meet its own needs — an activity that gained this rare tortoise its common name.

Removal of pine forests, land use changes and practices, along with encroaching urbanization have substantially reduced the Gopher Tortoises' preferred habitat. Remnant populations exist in the few areas still suitable for their survival.

Poor survival rates of hatchling turtles, low reproductive rates, collecting by reptile enthusiasts and automobile mortalities also dampen this species' chances for survival in the wild.

Gopher Tortoises were placed on the United States Endangered Species List in 1987 as threatened west of the Tombigbee and Mobile Rivers in Alabama, Mississippi, and Louisiana. They are also protected by Mississippi and Alabama state laws, and by CITES Appendix I.

While the Gopher Tortoise is not traded for its shell, leather or meat, it is collected and possibly traded among reptile collectors. Preservation of remaining Gopher Tortoise habitat is paramount to its survival, as are efforts aimed at reducing mortalities caused by automobiles. Continued monitoring is necessary.

HABITAT:
This long-lived reptile once inhabited all suitable habitats—sandy soils in pine forests—over much of the southeast and into Louisiana.

Gopher Tortoise *Gopherus polyphemus* **Guy Crittenden**

Audubon's Crested Caracara

Polyborus plancus audobonii
Appendix I

The official bird of Mexico, and the only Caracara species indigenous to the United States, this bird is one of the few raptors that both scavenges for dead animals and hunts living prey to fulfill its feeding requirements. To this end it physically resembles both vulture and eagle, and has been seen attacking and killing mammals, other birds and reptiles, as well as sitting and feasting with buzzards on rotting carrion.

Crested Caracara prefer wetland and prairie-type habitat. Development and alteration of those areas, which continue to shrink, is responsible for the decline of the Crested Caracara.

Audubon's Crested Caracara is protected under the U.S. Endangered Species Act and CITES Appendix I. There is no known illegal trade in the species.

Protection from development in the remaining Caracara habitat along with acquisition of additional suitable habitat remains the best option for this bird's survival.

HABITAT:
Once found in limited numbers in New Mexico, Texas, Arizona, Louisiana and Florida through Mexico into Central America, the species has disappeared in the U.S. in all states but Florida, where a few hundred reside in the central part of the state.

Audubon's Crested Caracara *Polyborus plancus audobonii* **Guy Crittenden**

Masked Bobwhite

Colinus virginianus ridgwayi
Appendix I

This beautifully marked quail once graced the grasslands of the Southwest, but loss of habitat eliminated the species from Arizona by the early 20th century.

Early attempts to restock the birds into New Mexico and Arizona failed, but subsequent attempts during the 1970s were successful on private ranch land containing suitable habitat.

This land was purchased by the Fish and Wildlife Service and has been established as the Buenos Aires National Wildlife Refuge. It is hoped that the species will successfully breed on the refuge establishing itself, once again, in Arizona.

Recovery efforts on the Buenos Aires Refuge have concentrated on habitat improvement and restocking efforts. Since 1985 approximately ten thousand Masked Bobwhites have been released with more releases planned. Field work in Mexico to monitor population and habitat changes is also a part of the recovery program.

Masked Bobwhite are protected by state law, the United States Endangered Species Act, and CITES. While the bobwhite's future is uncertain in Mexico, the species may be undergoing a comeback in Arizona after a century's hiatus.

HABITAT:
Desert grasslands, scrub brush and riparian valleys of south central Arizona and Sonoran Mexico constituted the original range of this species. But today Masked Bobwhite are restricted to the Sonoran grasslands of Mexico and the Buenos Aires National Wildlife Refuge in Arizona.

Masked Bobwhite *Colinus virginianus ridgwayi* Guy Crittenden

Great Horned Owl

Bubo virginianus
Appendix III

In its preferred forest habitat, the Great Horned Owl hears a chorus of sounds all but inaudible to the human ear, and its visual acuity in darkness is greater than human vision during daylight. These greatly enhanced visual and auditory senses make the Great Horned Owl the supreme nocturnal predator.

The Great Horned Owl has a wingspan of nearly five feet and may be up to twenty-five inches in length. Coloration may vary from dark brown to nearly black phases, to very light in the northern races. Its horns, or hair tufts, are thought to aid in camouflage, and distinguish the Great Horned from other large owls. With luck a Great Horned Owl may live twenty years.

The species is an opportunistic predator, and if its normal prey species are scarce, it may take domestic poultry or farm-raised game birds. While shooting owls is illegal, some Great Horned Owls are undoubtedly shot by farmers or gamekeepers. Other threats include collisions with cars and power lines, pesticides and loss of suitable habitat for nesting and hunting .

No known trade exists for Great Horned Owls and the species, as a whole, is thought to be relatively healthy.

The Great Horned Owl is highly beneficial to man as a rodent exterminator. One examined nest contained evidence of almost 120 rats that had been taken in a week. Great Horned Owls also help control disease outbreaks by pruning sick or dying rodents.

HABITAT:
The Great Horned Owl is found throughout North America with highest concentrations occurring where small mammals and rodents are plentiful. This species follows seasonal migration patterns closely tied with prey availability.

Great Horned Owl *Bubo virginianus* Wayne Dowdy

Mexican Bobcat

F. rufa escuinapae
Appendix I

The rugged habitat of the Mexican Bobcat requires that it be a more opportunistic predator than its northern cousins; it may kill and eat anything from reptiles to birds and mammals.

Three other bobcat subspecies occur in Mexico but no clear geographic boundaries separate the races. The four subspecies may overlap; their similar appearance makes it difficult to identify the subspecies.

The Mexican Bobcat is not currently threatened in Mexico; in fact it is considered to be widespread and abundant away from population centers. While flexible in its habitat requirements and quite adaptive to change, the range of this cat is shrinking because of agricultural development. Mexican Bobcats are hunted and trapped for local markets and the international fur trade. While poaching and illegal trapping occur, over-exploitation is not considered a threat to the species.

The bobcat is a major component in the international fur market. Illegal trade has increased as the bobcat pelt market has expanded. In 1989 one thousand bobcat furs were confiscated by the U.S. Fish and Wildlife Service in an effort to document the illegal trade of bobcat furs from protected Mexican species. However, the pelts are indistinguishable as to subspecies, making it next to impossible to determine type of cat or where it originated.

The species is protected under Mexican laws, which are often confusing and inconsistently enforced. Hunting is allowed in several Mexican states.

Biologists feel that the Mexican Bobcat's population is stable and not threatened. However, with ever-increasing habitat loss and the potential demands of the fur trade as threats, this animal needs close scrutiny.

HABITAT:
The Mexican Bobcat is found only in North Central Mexico, from the Tropic of Cancer south to Guerrero and along the West Coast.

Mexican Bobcat *F. rufa escuinapae* **Wayne Dowdy**

Northern Harrier

Circus cyaneus
Appendix III

More commonly known as the Marsh Hawk, harriers are typically viewed flying several feet above reeds or marsh grasses in search of rodents or birds. They can be seen throughout the day, but are found more commonly in the evening or early morning hours.

The only harrier species found in North America (there are ten worldwide), the Northern Harrier can be readily identified by its long slender wings and tail, and a prominent white rump patch. Harriers differ from most other birds of prey in that there are differences in plumage among adults of both sexes. With most birds of prey, the female is recognized by her larger size. The adult male harrier sports brighter and more obvious plumage, while the female is much duller hued. Being a ground nester, the female's less conspicuous markings probably serve her well as camouflage.

The Northern Harrier is protected through various raptor and migrating bird laws, and from commercial trade through CITES.

Protection of marsh land and open grassland habitat, fast disappearing in the United States, is critical to the Northern Harrier's future existence.

HABITAT:
The Northern Harrier is found from the western United States through the mid-latitudes into northeastern United States, and suitable habitats of Canada and Alaska.

Northern Harrier Circus cyaneus **Wayne Dowdy**

Mexican Bighorn Sheep

Ovis canadensis Mexicana
Appendix II

This incredibly hardy sheep has managed to adapt to the arid mountain range of south New Mexico and Chihuahua, Mexico, where drought is the norm. The Desert Bighorn adapts by acquiring all of its water from the various desert plants; it may actually never drink water throughout its life.

Also called the Desert Bighorn Sheep, this subspecies is somewhat smaller and more muscular looking than its northern counterparts. Desert Bighorn have shorter hair and slightly shorter horns, as well as a paler coat than their northern cousins.

All bighorn sheep are considered by trophy hunters to be the most challenging of hoofed mammals to hunt. This is especially true of the Desert Bighorn; only the most determined, in-shape hunter attempts to hunt this species. Sheep hunting is closed in Mexico. Several southwestern states offer very limited hunting on closely related bighorn species, but hunting is tightly controlled and not a threat.

Over-hunting by market and subsistence hunters, in addition to habitat changes caused by overgrazing and water diversion projects seriously depleted the Desert Bighorn populations. Protective measures, closing hunting seasons and establishment of sheep preserves have aided in the species' recovery.

The nearly inaccessible terrain of the Desert Bighorn's habitat makes poaching impractical, and it is not considered a serious threat. Reestablishing the species over its former range and protecting grazing habitat from domestic stock are essential components of the management plan for the Desert Bighorn's recovery.

HABITAT:
The Desert Bighorn is found in the mountains of southern New Mexico and Chihuahua, Mexico.

Mexican Bighorn Sheep *Ovis canadensis Mexicana* **Wayne Dowdy**

Scarlet Ibis

Eudocimus ruber
Appendix II

This gorgeous wading bird walks about in shallow waters, clapping its bill upon food items that include aquatic insects, crabs and snails. Found in flocks ranging from a dozen to several thousand birds or more, Scarlet Ibis were at one time the most common among the aquatic bird species of the Amazon region.

Today, the species has been nearly eliminated from much of this area, and has declined substantially throughout its range. In North America the occurrence of the Scarlet Ibis is accidental, but the bird may be found in Florida since the introduction of Scarlet Ibis eggs into White Ibis nests.

Scarlet Ibis were hunted for their brilliant crimson plumage, but the chief area of concern today is coastal pollution and destruction of mangrove-type habitats. Additionally, human intrusions into nesting areas will cause the birds to abandon their nest sites.

Preservation of Scarlet Ibis habitat from development and pollution and protection of nesting sanctuaries against human intrusion are of critical importance to this bird's survival.

HABITAT:
Scarlet Ibis occur in the lower Amazon river basin, in other northern South American countries and in Central America. The birds are found, rarely, in the Gulf Coast of Florida. The species has been released in Florida, and will breed with its northern race, the White Ibis. Habitat requirements include lagoons, mangrove forests and coastal swamps and tidewater rivers.

Scarlet Ibis *Eudocimus ruber* Wayne Dowdy

Jaguarundi

Felis yagouaroundi
Appendix I

Extreme shyness and avoidance of contact with humans have not protected the Jaguarundi from endangered status. Loss of habitat is the primary threat to this graceful animal's survival.

The cat's reclusive nature has prevented wildlife biologists from learning much about the animal, and the Jaguarundi's preference for dense mesquite thickets and nearly impenetrable brush makes field study very difficult.

It is thought that the Jaguarundi feeds on small rodents and ground nesting birds, and that it is a nocturnal hunter.

This species is not necessarily targeted by the fur trade but it does enter the black market as ocelot fur.

Jaguarundi are protected under the Endangered Species Act, CITES, and state laws. The worst peril facing the Jaguarundi is the conversion of brush-type habitat to agricultural and livestock needs.

The U.S. Fish and Wildlife Service plans to reintroduce Jaguarundi into suitable habitat, and a captive breeding program is underway at the University of Arizona. More observation and study efforts will be necessary to more fully understand the Jaguarundi's requirements in the wild.

HABITAT:
Southern Texas, southeastern Arizona, the coasts of Mexico and into Central America once were home to the Jaguarundi, but today it is no longer found in Arizona or southern Texas.

Jaguarundi *Felis yagouaroundi* Laura Mark-Finberg

Bald Eagle

Haliaeetus leucocephalus
Appendix I

Although it possesses great power and formidable weaponry, the Bald Eagle feeds primarily on fish and scavenged carrion. Often depicted as a symbol of ferocity and strength, it is not nearly as fearsome and active a predator as the only other eagle native to the United States and Canada, the Golden Eagle.

The Bald Eagle began declining during the 1800s because of deforestation, shooting, trapping, and the pressures of civilization. However, a more serious threat to the species entered the scene during the mid-20th century. DDT ingested by eagles through prey species caused the birds' eggshells to thin to the point where incubation became nearly impossible because the eggs would be crushed. Eagle numbers plummeted (as did those of many other raptor species).

The DDT ban signaled the beginning of the recovery of Bald Eagles and other raptors.

The species has made a remarkable comeback and is no longer endangered. But threats in the form of power line electrocution, accidental trapping, shooting, and lead pellet poisoning persist. Eagle feathers and talons are procured by Native Americans for use in traditional Indian ceremonies.

An underground trade in eagle parts has been documented but its scope is not known.

The recovery of the Bald Eagle is one of the most heralded conservation success stories of the 20th century. The species is protected by the Bald Eagle Protection Act and other state and national laws, and by CITES.

The mandatory change from lead shot to steel shot will reduce eagle mortalities from lead poisoning, and rewards for those turning in eagle poachers, such as the National Wildlife Federation offers, will keep shooting deaths down. Design changes in power line configuration to reduce death by electrocution, and a continued ban on DDT will enable the Bald Eagle to continue its majestic existence, and its status as the U.S. national symbol.

HABITAT:
Bald eagles once graced the skies of all North America and nested in nearly every state and throughout Canada.

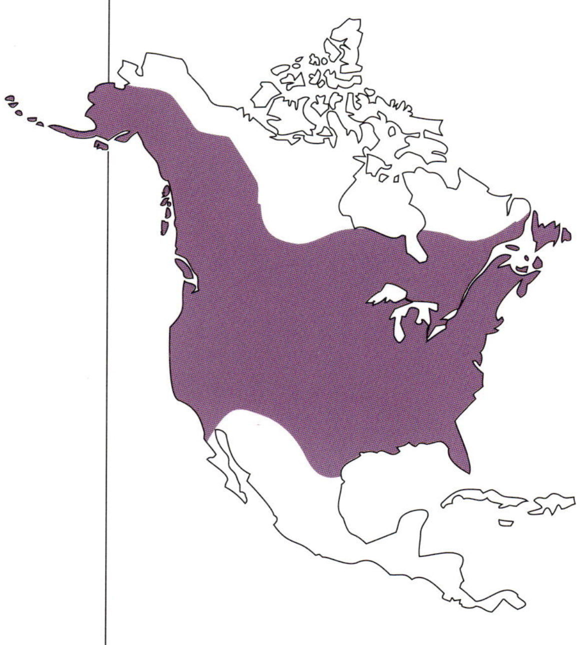

Bald Eagle *Haliaeetus leucocephalus* Laura Mark-Finberg

Peregrine Falcon

Falco peregrinus
Appendix I

Feeding primarily on other birds, the Peregrine Falcon relies on speed of wing to catch its prey, and, among falcons, is second only to the prairie falcon in swiftness.

The chief culprit in reducing falcon populations has been ingestion of pesticides, which has resulted in reducing or eliminating reproduction. The taking of Peregrines from the wild for falconry, while not necessarily a threat to strong populations, does pose a threat when levels are depleted.

The Peregrine's popularity among falconers gives it a high potential for trade. Illegal trade certainly exists, but the level of threat is unknown.

The Peregrine and three of its subspecies are protected under the Migratory Bird Treaty Act of 1975, the Endangered Species Act, and by many state wildlife laws and regulations. It is an Appendix I CITES listed species, affording it the highest protection from trade.

The elimination of pesticides — especially DDT — worldwide would be an important step in the recovery of Peregrines (and all raptors, for that matter). Curtailing the capture of wild specimens for the falconry trade is also an important element to this bird's survival.

Restocking programs now play a role in expending the Peregrine's range and numbers.

HABITAT:
Peregrines range worldwide, but have been eliminated, or severely depleted, throughout much of their range. They are found throughout the United States with the exception of the north central region. Peregrines breed primarily in Canada.

Peregrin Falcon *Falco peregrinus* Laura Mark-Finberg

American Kestrel

Falco sparverius
Appendix II

The Kestrel, previously known as the sparrow hawk, has the distinction of being the smallest and most numerous raptor in North America. Primarily an insect eater, it is known to dine on rodents and reptiles on occasion.

Kestrel populations have remained stable and even increased over the years because of the bird's diet. A menu comprised largely of insects keeps this falcon relatively free of the pesticide problems faced by other raptors.

Additionally, Kestrels prefer a more open habitat. Loss of forests has actually benefited the species. The American Kestrel is second in popularity to the red tailed hawk among falconers, but the capture of wild specimens for the sport does not pose a threat at this time. Because it eats insects, it is used primarily by novice falconers.

In addition to CITES protection, Kestrels are protected under United States fish and wildlife laws, Canadian laws, and state laws.

No serious trade, either nationally or internationally, exists for the species. Kestrels seem to thrive in close proximity to man and agricultural practices. Given an abundant food supply and few enemies, it is one of the few species of raptors whose future seems to be unclouded.

HABITAT:
The American Kestrel is found throughout Canada and the United States, through Central America and into South America.

American Kestrel *Falco sparverius* **Laura Mark-Finberg**

Northern Goshawk

Accipiter gentilis
Appendix II

One of the largest of the hawk species, the Goshawk is readily capable of taking prey as large as the snowshoe hare.

It is thought that because Goshawk habitat is far from human population and agricultural practices, they have not been nearly as affected by pesticides as have other birds of prey. Goshawks are currently a stable species, and in some states, such as Alaska, the Goshawk population is high enough to allow its use among falconers.

Since Goshawk habitat is generally unsuited for logging and agriculture, habitat destruction is not at this time posing a serious threat to the species.

Goshawks are protected under the U.S. Migratory Bird Treaty Act, state laws and regulations, and CITES. The species' desirability among falconers worldwide creates a black market trade of unknown proportions. This hawk is highly esteemed by falconers in North America, Europe, and especially Asia.

Due to its popularity, the Goshawk could be needlessly exploited, resulting in a downturn of its population trend. Continued protection and monitoring by CITES will help keep this situation from occurring.

HABITAT:
This species occurs primarily in the northern hemisphere taiga regions and in high mountain regions to Mexico.

Northern Goshawk *Accipiter gentilis* Laura Mark-Finberg

Harpy Eagle

Harpia harpyja
Appendix I

Known as the "King of Eagles," Harpy Eagles weigh up to twenty pounds, and are the largest eagles in the world.

This terrifying hunter of the tropical rain forest is equipped with huge talons the size of grizzly bear claws — the largest claws in the animal world. These formidable weapons are wielded with tremendous force upon sloths, howler monkeys, birds, reptiles and other arboreal prey species.

Harpys are generally considered to be endangered throughout their range. Loss of habitat, primarily due to logging of the large trees this eagle requires for nesting, is the principle cause of its decline. Harpy Eagles probably never were numerous because of the large territories they required for hunting.

The Harpy Eagle, the national bird of Panama, is protected in most countries which have populations of the species. An important conservation project is the Harpy Eagle Conservation Program, which intends to establish or strengthen conservation projects in all countries harboring Harpy Eagles.

Harpy Eagles were given Appendix I CITES protection in 1975. Population numbers are currently unknown.

HABITAT:
The range of this rare eagle is believed to extend from southern Mexico through central America to northern Argentina.

Harpy Eagle *Harpia harpyja* Laura Mark-Finberg

American Ginseng

Panax quinquefolium
Appendix II

The roots of American Ginseng are in high demand, especially from the Orient, where it is used for medical and cooking purposes, as well as its supposed qualities as an aphrodisiac. While there is intensive cultivation of ginseng, wild ginseng root is much more highly valued than its field-cultivated counterpart. The value of wild collected and cultivated ginseng root is at least twenty-five million dollars, and growing.

Over-harvesting of wild ginseng is a concern and threatens to greatly diminish the species over much of its native habitat. Also of concern is loss of habitat and logging, which removes the forest canopy essential to the plant's existence.

American Ginseng is protected in most states where it is found and is protected under federal law as well.

American Ginseng has been afforded CITES Appendix II protection since 1975, but despite this and other protective measures, the species is declining in the wild.

HABITAT:
American Ginseng is a perennial herb which grows wild in shaded, loamy soils in the eastern, western and midwestern United States and Canada.

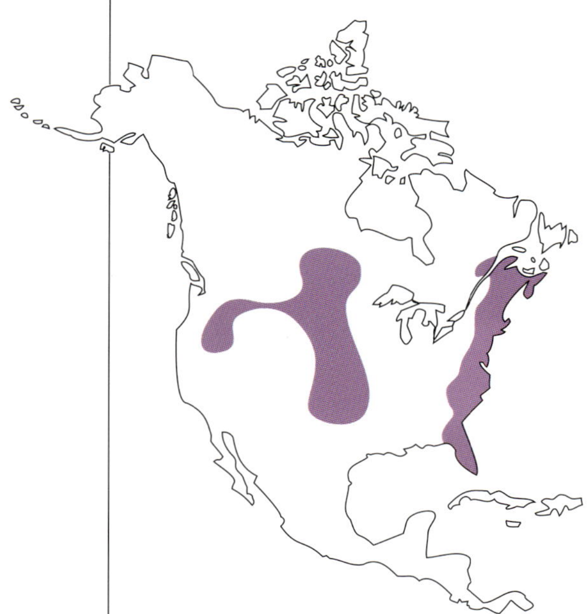

American Ginseng *Panax quinquefolium* Norma Hodge

Siskiyou Lewisia

Lewisia cotyledon
Appendix II

Siskiyou Lewisia flowers in April and May, and its attractive blooms and crisp-looking leaves make it a tempting plant to collect from the wild for one's garden. Collecting for horticultural purposes, recreational activity in the plant's habitat, and possibly mining are the primary threats faced by Siskiyou Lewisia.

A member of the purslane family, it exhibits several varietal forms and will hybridize with other types of cotyledon.

As well as being listed in the Endangered Species Act and CITES Appendix II, the plant is protected by various state laws.

It has been recommended that recreational activities in Siskiyou Lewisia habitat be closely monitored and, if necessary to protect sensitive populations, restricted. Additional protection may be necessary if mining in this species habitat is undertaken. Other cotyledon varieties are fairly common, especially in a few northern California and southern Oregon wilderness areas.

HABITAT:
Siskiyou Lewisia is endemic to wilderness areas of extreme northern California and southwest Oregon. Habitat preference is usually granite or serpentine substrate in oak-conifer forests.

Siskiyou Lewisia Lewisia cotyledon **Norma Hodge**

Onconee Bells

Shortia galacifolia
Appendix II

Onconee Bells is the only species of its kind (shortia) found in North America.

Inundation of much of the Onconee Bells' habitat by stream damming in the southern United States has severely depleted the plant's available growing areas, but where suitable habitat still exists the species is abundant.

Onconee Bells' habitat also has suffered from extensive logging and development, further limiting its range. The plant is gone from a high percentage of its previous territories. Fortunately, Onconee Bells is apparently fairly easy to cultivate. The plant has been reported from several other eastern and northern states, but these were cultivated plants rather than naturally-occurring specimens.

Onconee Bells are listed as endangered in Georgia, North and South Carolina. However, the protected land status of several large populations of the species has made additional protection unnecessary at this time. Onconee Bells are under consideration for federal listing as well as state designation for endangered species status.

HABITAT:
Habitat preference for Onconee Bells generally leans toward cool canyon-type drainages of northeastern Georgia, South and North Carolina.

Onconee Bells Shortia galacifolia **Norma Hodge**

Calypso Orchid

(also known as Fairy Slipper) C. bulbosa
Appendix II

This exquisite and delicate orchid family member is generally found in moist, mountain-type conifer forests in lower to middle elevations.

The Fairy Slipper, or Calypso Orchid, is considered by botanists to be a fairly common species throughout much of its range. It has been given CITES protection — as have all wild orchids — to protect endangered or threatened look-alike orchid species. Fairy Slippers are currently not threatened. However, logging poses a threat to regional populations, as does any development. Collecting does not pose a threat; the orchid will not survive being picked, and only amateurs will gather them.

HABITAT:
The Calypso Orchid grows over a wide geographic range, from Alaska to Labrador down the west coast of the United States into Arizona and New Mexico, and throughout the Great Lake states.

Calypso Orchid (Fairy Slipper)   C. bulbosa   Norma Hodge

California Pitcher Plant

(also known as Western Pitcher Plant or Cobra Lily) Darlingtonia californica
Appendix II

Formerly known as the Cobra Lily because of its ominous-looking purplish curved hood, the Western Pitcher Plant grows in acidic soils which offer little in the way of nutrients. The Western Pitcher Plant makes up for this nutrition loss by capturing insects, which fly into its hood and slide irreversibly down the plant's throat to their demise. The plant's prey are then broken down for nutrition by way of bacterial action, as it produces no digestive enzymes.

Bogs and seeps with running water and little nitrogen provide habitat for this rare carnivorous plant.

The Western Pitcher Plant faces a wide variety of threats, chief among which are over-collecting, wetland draining, logging and almost any development. Collecting poses a very serious threat even in protected areas. Horticultural collectors are particularly fond of carnivorous plants and have little difficulty finding illegal markets for their sale. Western Pitcher Plant populations are expected to plummet despite protective efforts and protected reserves.

HABITAT:
The coasts of central Oregon through southern Oregon into northern California are home to the majority of the plants, but a few may be found in the coastal mountains, as well. Western Pitcher Plant habitat is limited to heavily forested areas in watery, acidic soils.

California Pitcher Plant *Darlingtonia californica* Norma Hodge

Black Bear

Ursus americanus
Appendix II

With a population of 670,000 animals, the American Black Bear is not threatened or endangered in the United States or Canada — with the exception of the Louisiana Black Bear population, which is proposed for endangered species listing by the U.S. Fish and Wildlife Service.

Its listing in Appendix II of CITES is highly controversial, but the CITES listing is justified by the Black Bear's look-alike status with other bear species which are threatened or endangered.

CITES has found great difficulty in differentiating bear parts and derivatives, and this has handicapped efforts to effectively control the trade of these items, especially gall bladders and paws; bear parts across species are indistinguishable.

This trade is threatening the very survival of Appendix I bear species. The commercial value of bear gall bladders and paws (prized for alleged aphrodisiacal properties) is increasing. This demand is beginning to exert pressure on the Black Bear.

Whole Black Bear carcasses are being found with nothing but gall bladders and claws removed. Many prosecutions for illegal sales of Black Bear parts are connected to the Asian trade.

But the primary reason for listing the Black Bear is to protect the endangered Asian bear species which are being illegally traded.

The Black Bear is well managed in Canada and the United States as a big game animal, and is in no danger at this point. Protection by CITES seems desirable not only to protect the Asian bear but to keep the Black Bear from becoming threatened by illegal poaching aimed at providing bear parts to the highly lucrative gall bladder markets.

HABITAT:
The Black Bear is distributed from Alaska through Canada and in thirty-two of the United States.

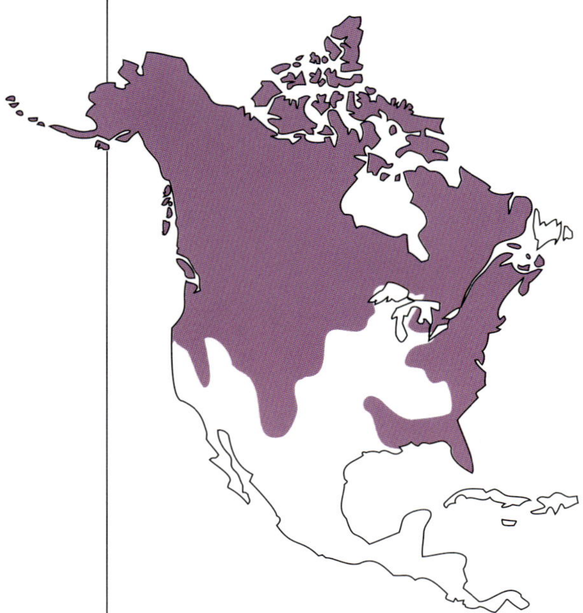

Black Bear *Ursus americanus* Laney

Osprey

Pandion haliaeetus
Appendix III

Among the most widespread of all raptors, it is the Osprey's talon structure that places the bird in a family all its own. The Osprey's uniquely adapted talons are formed to grasp and hold its principle prey species: fish.

While perched, the Osprey is quite eagle-like in appearance, but in flight it appears more gull-like, but with a larger, four and one-half foot wingspan.

Ospreys are almost always found around large lakes, rivers, or estuaries where high populations of fish can be exploited. When hunting, the Osprey cruises at one hundred or so feet as it looks for surface feeding or lounging fish.

Once the Osprey spots a likely meal it hovers for a few seconds, folds its wings, and plummets at amazing speed toward its prey. Rising somewhat clumsily from the water, the bird shivers off excess moisture and arranges the fish head first in its talons for a more aerodynamic flight back to its nest or perch.

Ospreys were in serious decline over all of North America before a ban on egg-thinning DDT pesticides helped the species make a comeback. Shooting and egg collecting (the Osprey's eggs were highly coveted by collectors) were also threatening the bird's future. However, the species has made a very strong comeback.

Osprey are protected by many national and state laws, as well as CITES. No known trade exists; however some birds are taken for their talons and feathers for ornamentation and ceremonial purposes.

HABITAT:
The Osprey is found on all continents and on many islands. In North America they occur in the Pacific and Rocky Mountains states, the Atlantic states through Florida and over much of Canada and Alaska. Nearly the entire North American population migrates to Central and South America to over winter.

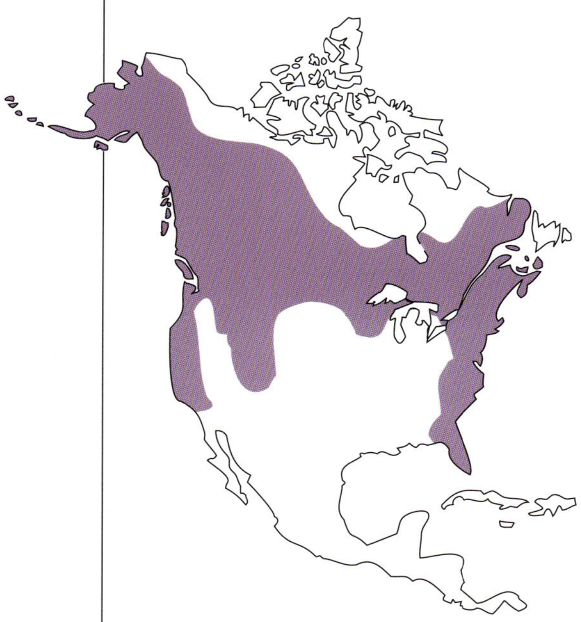

Osprey Pandion haliaeetus **Laney**

Red-Tailed Hawk

Buteo jamaicensis
Appendix III

The Red-Tailed Hawk prefers open sparse stands of woods and wooded lots adjacent to fields, pastures, and meadows. They can frequently be seen perched upon tree branches, fence posts, or other structures as they keep a vigil for rodents emerging from vegetation, birds feeding along brush lines, or other small, unwary prey. Red-Tailed Hawks also will make a meal of insects, snakes, and, occasionally, fowl such as pheasants and chicken. Any damage the hawk may do to domestic fowl is more than offset by the rodents and other "pest" animals they take. Among falconers the Red-Tailed Hawk is number one in popularity.

This large hawk of the Buteo family has several subspecies with varying colors and patterns. In general, most Red-Tailed Hawk "races" have a wingspan of four feet or more, a length of seventeen to twenty-two inches, and weigh about three and a half pounds. Size varies somewhat according to sex and race.

Red-Tailed Hawk numbers have rebounded since the ban on DDT and the status change from vermin to protected bird of prey.

In addition to CITES trade protection, the species is protected by state and federal laws, Canadian laws, and various international laws. Its numbers may still be increasing and its range is believed to be expanding. Continued efforts aimed at raptor protection and population monitoring should allow Red Tails to flourish.

HABITAT:
Probably the most numerous and widespread of the bigger hawk species, Red-Tailed Hawks are common throughout the United States and Canada, extending into Alaska with subspecies found in Mexico. As with most hawks, Red Tails will migrate from areas of depleted prey species into territories promising better hunting.

Red Tailed Hawk   *Buteo jamaicensis*   **Jan Martin McGuire**

Aleutian Canada Goose

Branta canadensis leucopareia
Appendix I

When Arctic foxes were introduced to the Aleutian Island chain during the 1800s, the Aleutian goose found itself with an enemy that threatened its survival as a species.

Introduced for the fur trade, the foxes decimated goose populations by destroying nests, dining on eggs and catching and eating goslings.

Attempts to reestablish this species over several of the islands have been successful in large part because of fox eradication programs. Other efforts to bring the species back included closing goose hunting in areas where the bird winters in central California. Aleutian Canada Geese are similar in appearance to many other Canada goose subspecies, and some birds are killed by hunters despite hunting closures.

The subspecies is protected through the United States Endangered Species Act, where it was recently down-listed to threatened.

In addition to population increases, the Migratory Bird Treaty Act, state laws, protected refuges, and CITES Appendix I listing are helping ensure the species' survival. The delisting from endangered to threatened by the United States Endangered Species Act indicates that biologists are confident that the bird is well on the road to recovery.

With constant study and monitoring of population trends, continued regulatory and refuge protection, the Aleutian Canada Goose should fare well.

HABITAT:
The historical range of the Aleutian Canada Goose included the Soviet Union, Japan, Canada, Mexico, and the United States. Current breeding populations are confined to a handful of islands in the Aleutian Island chain. Migration takes this species as far south as the central valley of California and possibly Mexico.

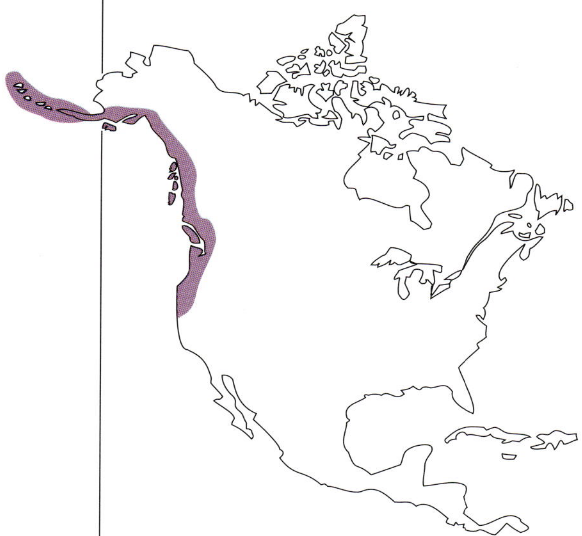

Aleutian Canada Goose *Branta canadensis leucopareia* **Jan Martin McGuire**

Polar Bear

U. thalarctos maritimus
Appendix II

The largest of all land carnivores, the Polar Bear may reach fifteen hundred pounds. Only the brown bear rivals it in size. The Polar Bear survives in the inhospitable arctic climate with the aid of a thick layer of blubber, hollow insulating fur and hibernation during the sub-zero arctic winters.

Polar Bear populations are relatively stable, but continued development of Polar Bear habitat, especially for oil and gas exploration and extraction, pose a very real danger to the species.

While Polar Bear are hunted by native peoples for subsistence and ceremonial purposes, the take is strictly controlled and relatively minimal.

With the exception of trophy hunting, there is no legal international trade in Polar Bear, but a black market for Polar Bear parts does exist.

Polar Bear are afforded protection internationally through CITES and a five-country agreement which restricts harvest to native subsistence or scientific research. They are additionally protected under the Marine Mammal Protection Act of 1972, Alaskan state wildlife laws, and Canadian laws.

Protection of arctic areas where Polar Bear habitat occurs is critically important to the species' survival. Close monitoring of potentially illegal trade should continue.

HABITAT:
The Polar Bear is endemic throughout the arctic regions, including parts of Alaska. It is also found in James and Hudson Bays in Canada. Occasionally Polar Bear are found far out in the ocean on ice floes.

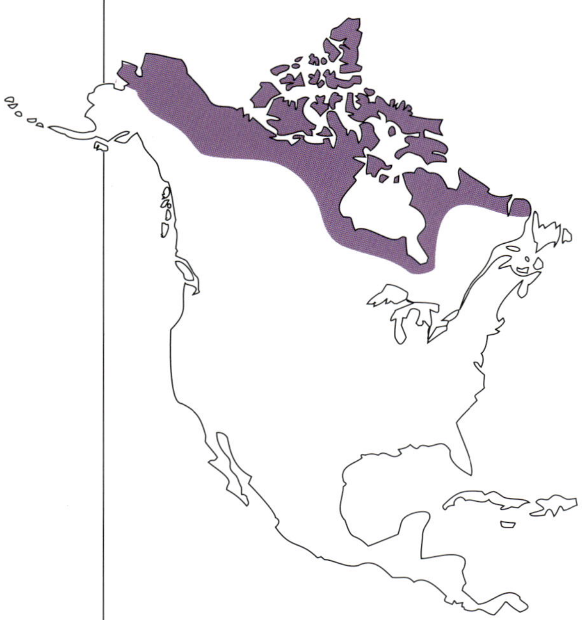

Polar Bear U. thalarctos maritimus **Jan Martin McGuire**

Gyrfalcon

Falco rusticolis
Appendix I

With a wingspan reaching four feet, the Gyrfalcon is the largest of all the falcon species, and the only falcon exhibiting three color phases: gray, white and black.

The most northerly occurring of all the hawks and falcons, the Gyrfalcon is rarely found south of the Arctic.

In years of low prey numbers the high arctic tundra habitat is extremely fragile. Protection of this habitat is critical to the Gyrfalcon's survival.

While their generally remote living conditions make the Gyrfalcon difficult to obtain, they are nevertheless highly desirable specimens for the falcon trade. Regulations forbid their wild capture.

A strong black market potential exists for this species on a world wide basis.

Gyrfalcon are protected under CITES Appendix I, migratory bird treaties, and laws of states and provinces. Tundra protection and close monitoring of the trade in falcons as well as DDT and other pesticides which cause thinning of eggshells are imperative.

HABITAT:
While usually found in the high Arctic, the Gyrfalcon may winter over as far south as the United States - Canadian border. It may also migrate south from the arctic tundra habitat in years of low prey (lemming and hare) populations.

Gyrfalcon *Falco rusticolis* **Jan Martin McGuire**

River Otter

Lutra canadensis
Appendix II

Most members of the weasel family have reputations as ferocious, aggressive, even mean-spirited predators. But not the fun-loving River Otter, who revels in the offering of a mud slide, or in a game of tag with a trout. Otters are genuinely playful and spend most of their awake hours in pursuit of games. Along with a playful nature, the River Otter's chattering and chuckles seem to completely complement its personality.

The River Otter's elongated body may reach thirty inches in length, with a twelve- to eighteen-inch heavily furred tail. They weigh from fifteen to thirty pounds. These riverine carnivores have beautiful, rich, brown upper coats with silvery bellies. Their coats were highly desired by the fur trade and over-trapping led to serious population declines in the past.

Favored haunts are rivers, lakes, and swamps where fish, amphibians, or crayfish are abundant. Otters often travel and hunt in family groups, but even serious hunting forays often turn into high jinks.

Now protected from trapping, otter are again appearing in lakes and rivers after a long absence. Their range appears to be spreading in some Pacific states.

HABITAT:
The River Otter is found in Alaska, through Canada and over most of the United States. They do not hibernate and must hunt year round to survive.

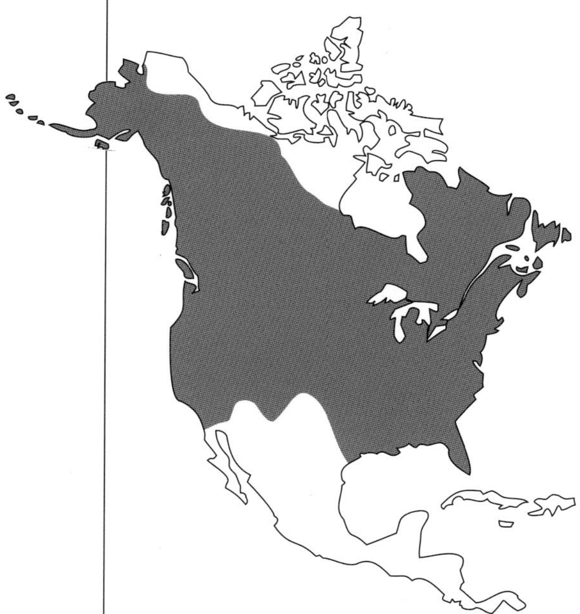

River Otter *Lutra canadensis* Jan Martin McGuire

Golden Eagle

Aquila chrysaetos
Appendix II

HABITAT:
The Golden Eagle is found in highest concentrations in the western United States and Canada, especially in the Rocky Mountain states, eastern British Columbia, and Alberta. It has been sighted during winter months throughout most of Canada and the United States, as far north as Alaska, and as far south as Florida.

Seeing this huge raptor soaring in thermal updrafts or diving at unbelievable speeds toward an intended meal is one of nature's most awe-inspiring visual treats.

This bird's name — 'Aquila', Latin for eagle, and 'chrysos', Greek for golden — is apt for identification purposes; its golden or yellowish colored nape stays the same color from immature young through adult, and is an easy identifier for North America's largest eagle.

The Golden Eagle has a reputation as a ferocious hunter. With a wingspan of seven or eight feet and weighing up to fourteen pounds, this eagle is indeed capable of taking on large prey species. Known to be included in its diet are pronghorn antelope, deer, sheep, jackrabbit, and large birds such as geese and cranes. Even other predators such as coyotes, foxes, and badgers have fallen victims to this bird's grasping talons.

Being a very agile flier for its size is a needed advantage for taking on prey as quick as the pronghorn antelope. This master of the air can dive at speeds approaching two hundred miles per hour.

Golden Eagles will also take advantage of carrion such as dead sheep, and as a result often are wrongfully blamed for the death of domestic animals. While they are capable of taking sheep, it is extremely rare for them to do so. Those rare instances have, nevertheless, cost the lives of thousands of eagles and other predator species through trapping, poisoning, shooting and other predator control programs.

The Golden Eagle is now protected under numerous state and federal laws and CITES, but some are still shot, poisoned, or trapped by ranchers erroneously assuming they are a threat to livestock. Despite this, the Golden Eagle's population is stable.

Golden Eagle *Aquila chrysaetos* Jan Martin McGuire

Hawaiian Hawk

Buteo solitarius
Appendix II

Known as the "Io" to the islanders, the Hawaiian Hawk is Hawaii's only native bird of prey. This darkish colored hawk is found from sea level to the mountains. It prefers forest habitat, but is adaptable to agricultural land if trees are present.

Hawaiian Hawk numbers have rebounded; the bird's main stronghold—Hawaiian Volcanoes National Park—records the population at about two thousand.

Loss of forest habitat, shooting, and nesting disruptions were the main causes for the species' decline, but the bird is now considered to be stable, especially within the park.

The Hawaiian Hawk is listed as threatened on the United States Endangered Species list, protected by Hawaiian state wildlife laws, and protected from illegal trade through CITES. There is currently no known trade on the species nor any black market.

The primary survival concern for the Hawaiian Hawk is preservation of its remaining habitat and disturbance of nest sites. Nests that are frequently disturbed are often unsuccessful in incubation of eggs or rearing of young. Nesting sites must be protected from disturbances if the Hawaiian Hawk is to flourish.

HABITAT:
The species nests only on the island of Hawaii, but has been sighted on several other islands.

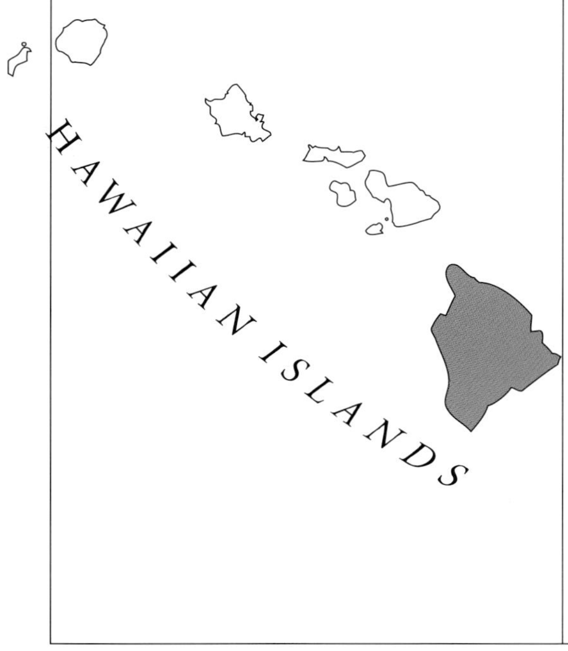

Hawaiian Hawk *Buteo solitarius* Lawrence B. McQueen

Laysan Duck

(also known as Laysan Teal) Anas laysanensis
Appendix I

In 1930, only one wild female of this species, found on Laysan Island off the Leeward Islands in the Hawaiian Archipelagos, was known to exist. Today, several hundred wild birds — descended from that one remaining female — endure, and several hundred more can be found in captive populations.

The main threat to Laysan Ducks remains the introduction of non-endemic species. Rabbits introduced to Laysan Island removed vital plant cover, bringing about the near-extinction of the ducks.

The rabbits are gone now, and the vegetation has recovered, but potential problems from accidental introductions are an ever-present menace. The restricted range and low population of the Laysan Duck make survival a delicate balance.

More Laysan Ducks exist in captivity than in the wild, and trade in wild specimens is a potential threat. In addition to protecting this species from trade, CITES has provided a forum for discussing how captive Laysan Ducks could be used for possible restocking, should the wild population fail.

This species is also protected under the U.S. Endangered Species Act.

The Laysan Duck and its habitat must be closely monitored and protected to ensure its survival.

HABITAT:
This species' geographic range is limited strictly to Laysan Island off the Leeward Islands in the Hawaiian Archipelagos. Laysan Island is part of the Hawaiian Island National Wildlife Refuge and is closed to entry except by permit.

Laysan Duck *Anas laysanensis* Lawrence B. McQueen

Eskimo Curlew

Numenius borealis
Appendix I

This long-legged bird with the thin down-curved bill is the smallest representative of the Curlew family. Also known as the Prairie Pigeon, the species was once thought to be extinct, but the Eskimo Curlew has been recently sighted in minimal numbers in Texas and northern Canada.

The total population is estimated at about one hundred fifty birds. The Curlew's rarity and its propensity toward isolation from civilization has limited studies of the bird. It is unclear as to what led to this species' decline, or how much it has declined from historical highs, but wetland drainage and alteration doubtless has had an effect. Market hunting during the 1800s probably also led to the species' decline.

Eskimo Curlew are protected under the U.S. Migratory Bird Act, the Endangered Species Act, Canadian wildlife laws, and CITES Appendix I.

No known black market trade exists. The American Ornithologists Union has recommended a management program aimed at protection of the Eskimo Curlew's migration routes and Curlew look-alike awareness programs.

HABITAT:
The Eskimo Curlew was found in northwestern Canada, Alaska, through the midwestern United States, into Texas, Louisiana, and Mississippi. Summer migrations occurred into Argentina.

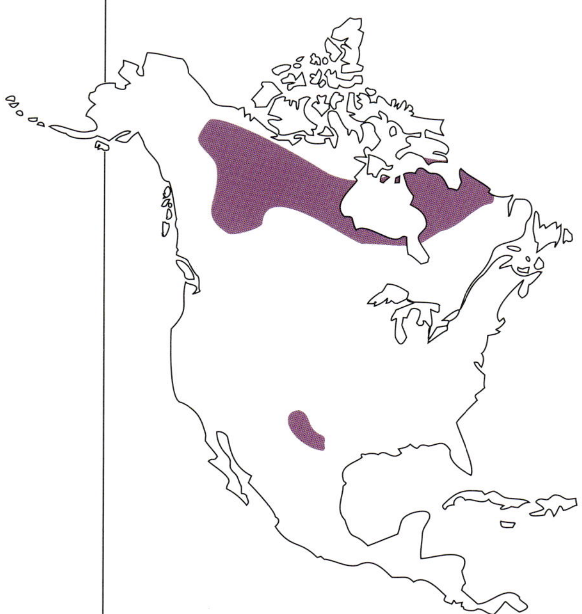

Eskimo Curlew *Numenius borealis* Lawrence B. McQueen

American Flamingo

Phoenicopterus ruber ruber
Appendix II

Considered the second rarest of all flamingo species, the American Flamingo is the most popular for zoos and bird collectors. In addition to the capture of wild specimens for the aviculture trade, flamingos are still hunted for food in some locales. The capture of wild flamingos for export to zoos and other collections constitutes the biggest threat to the species, as many captured birds are killed during capture or die during transit.

Also known as the Caribbean, Cuban, Rosy, or West Indian Flamingo, the American Flamingo is characterized by neon bright crimson feathering.

American Flamingos are protected by laws of the various countries in which they are found, but enforcement often is nil. Law enforcement needs to be vastly improved in some countries to protect the species from hunting and wild capture.

Additionally, the American Flamingo's habitat is fragile and easily disturbed. Many former nest and breeding areas have been abandoned, and close habitat monitoring is crucial.

HABITAT:
American Flamingo are restricted to islands or coasts which offer soda or salt-type lakes and marshes. The Caribbean, Galapagos Islands, Bahamas, Cuba, and the Yucatan Peninsula of Mexico constitute the range of this species.

American Flamingo *Phoenicopterus ruber ruber* Lawrence B. McQueen

Montezuma or Harlequin Quail

Cyrtonyx montezumae (Mexican population)
Appendix II

Often referred to by locals as "silly chicken" or "fool's quail," the Harlequin's docile nature makes it highly vulnerable to hunting by even the most primitive of methods.

In addition, habitat loss due to agricultural takeover and subdivision conversion poses a serious threat to the future of this bird. Its numbers are also depressed due to predation by natural and especially domesticated predators. Being a ground nester, the Harlequin loses many nests and broods to predators such as skunks, dogs and cats.

This smallest North American quail features a distinctly "Harlequin"-marked black and white face, a speckled chest, and a pale head crest. The female is duller hued. The unusually marked male Harlequin cannot be confused with any other quail species.

Protection of remaining habitat along with control of domestic predators are essential steps toward keeping this quail's viability as a species. Captive breeding and subsequent release back into suitable protected habitat may also play a role. Harlequin are protected by Arizona and Texas wildlife laws, and by CITES.

HABITAT:
The Harlequin's native habitat includes sparsely wooded mountains, grassland hills and canyons from southern Arizona into western Texas and Mexico.

Montezuma or Harlequin Quail Cyrtonyx montezumae **Lawrence B. McQueen**

Short-Tailed Albatross

Diomedea albatrus
Appendix I

Short-Tailed Albatross spend most of their life at sea riding the air currents and feeding on fish and squid. They only venture onto land, primarily volcanic islands, for breeding and nesting purposes. Albatross literally cannot fly, and rely on air currents to allow them to navigate the ocean skies. An albatross may spend as many as seven years at sea, touching neither land nor water, and never having flapped a wing.

Short-Tailed Albatross, called Ahodori ("stupid bird") in Japanese, were easily killed by Japanese hunters for the feather trade. Tens of millions were slaughtered during the 1800s into the twentieth century and, coupled with volcanic eruptions, which ruined vast areas of nesting grounds, the species was reduced to approximately fifty by 1939.

Short-Tailed Albatross had been presumed extinct after further volcanic eruptions seemed to have killed off the remaining birds. However, in 1950, ten Short-Tailed Albatross were sighted. Due to an end to hunting and protected sanctuaries the species has increased to perhaps several hundred birds.

HABITAT:
This rarest of all the albatross species was once found in abundance throughout the North Pacific, Japan and North American coasts. It is gone now from most of its original range, breeding on only a few Japanese islands.

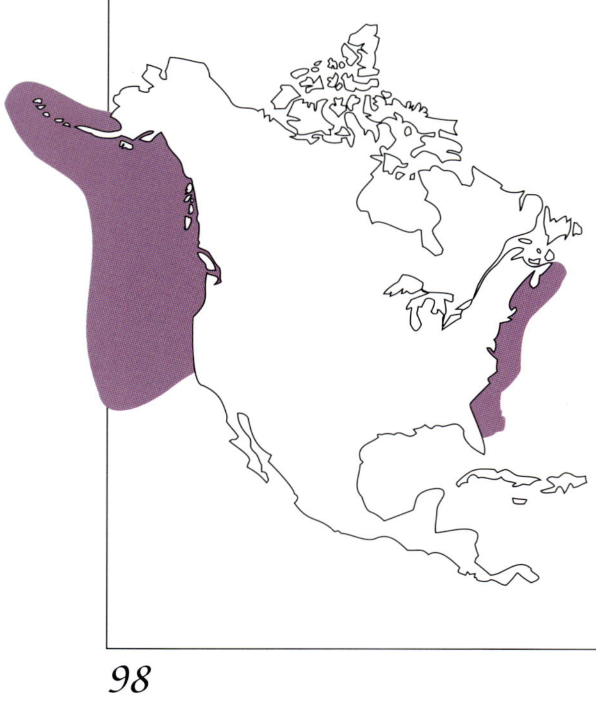

Short-Tailed Albatross *Diomedea albatrus* Lawrence B. McQueen

Queen Conch

Strombus gigas
Appendix I

Highly prized as a food delicacy, Queen Conch have become rare wherever divers can easily exploit them, but they may be numerous in less accessible waters. The Queen Conch's preferred habitat is grass beds near coral reefs or islands in relatively shallow water, where it feeds upon algae and sea grass.

The United States imports hundreds of metric tons of conch from the Caribbean, but the traditional U.S. fishery is gone. Conch is prized for its high value as food and for its value in the shell trade. The Queen Conch shell is one of the most popular species traded.

Stocks of Queen Conches are heavily depleted throughout their range, and their deeper water havens are now being scoured by scuba divers.

The species is protected in Florida through a moratorium on commercial fishing and a daily bag limit on recreational catches. The conch is also protected by international protection treaties, and CITES. Most countries now exporting Queen Conch recognize the need for tighter controls and restrictions for stocks to recover.

HABITAT:
Queen Conch are found in southeast Florida, throughout the Caribbean and in the southern Gulf of Mexico.

Queen Conch *Strombus gigas* Ron Pittard

Short-Nosed Sturgeon

Acipenser Brevirostrum
Appendix I

Sturgeon are considered the heavyweights of the freshwater fish world, but not this relatively small species. Unfortunately, size isn't everything, and during the 1800s commercial fishermen targeted the Short-Nosed Sturgeon for its eggs and meat, which they actually preferred to that of its larger cousin, the Atlantic Sturgeon.

Because there are so few Short-Nosed Sturgeon, the catch is incidental now.

While over-fishing resulted in its precipitous decline, dams, pollution, and dredging also contributed to deteriorating stocks of sturgeon.

At this time, there is no known trade on this species and little black marketing.

Short-Nosed Sturgeon are protected under Canadian and United States endangered species laws, state laws, and CITES.

Pollution-free waters, trade monitoring and protection from incidental gill netting continue to be the most important contributions to the survival of this sturgeon. The Hudson Bay region has reported increasing Short-Nosed Sturgeon populations, an encouraging sign.

HABITAT:
The Short-Nosed Sturgeon is found in estuaries and river systems — as well as offshore — along the entire eastern seaboard of North America.

Short-nosed Sturgeon *Acipenser brevirostrum* **Ron Pittard**

Atlantic Sturgeon

Acipenser oxyrhynchus
Appendix I

Atlantic Sturgeon have a tremendously high commercial value, both for their eggs, sold as caviar, and their flesh. In addition to their commercial value, an unknown percentage are killed in fishing nets.

Hundreds of tons of Atlantic Sturgeon eggs were shipped to primarily European markets during the late 1960s, but commercial shipments are far fewer today. Pollution has also negatively affected the populations. Smoked sturgeon meat is highly prized and makes up a large legal and illegal commercial trade in China, Europe, Japan, and Asia.

In addition to CITES Appendix I listing, Atlantic Sturgeon are protected under federal, state and Canadian laws.

The Atlantic Sturgeon is one of several look-alike species of sturgeon, at least one other of which is also CITES listed.

Atlantic Sturgeon spawn in freshwater rivers and young are reared in the river-estuarine system. Because they spend so much of their lives in the influences of river systems, this species is highly susceptible to industrial pollution.

When that susceptibility is coupled with their highly esteemed commercial products—caviar and smoked sturgeon meat—it is apparent why the Atlantic Sturgeon requires careful monitoring and intense management to ensure its survival as a commercial entity.

HABITAT:
Ranging from Nova Scotia and New Brunswick down the Atlantic Coast to eastern Florida, the anadromous Atlantic Sturgeon may be found in river systems, estuarine areas, and/or salt water.

Atlantic Sturgeon *Acipenser oxyrhynchus* **Ron Pittard**

Paddlefish

Polyodon spathula
Appendix II

Dams, river channelization, pollution, and over-fishing have significantly reduced the populations and range of this bottom-dwelling scavenger, also known as the Spoonbill. Currently, while populations are down from historical highs, the species as a whole is not rare, nor — with some localized exceptions — endangered.

Paddlefish roe, illegally sold as lumpfish eggs, is commercially sold to United States, Japanese and European markets.

Some states allow limited sport fishing or snagging for Paddlefish, but this does not pose a threat. Captive breeding for stocking and the aquaria trade is ongoing. The roe is highly prized, bringing hundreds or even thousands of dollars per pound at retail from some United States dealers.

Biologists are concerned that existing state laws may not sufficiently regulate the trade and reduce the illegal catch of this species.

A CITES Appendix II Listing will protect Paddlefish from potential excessive harvest caused by the high demand for caviar.

HABITAT:
The Paddlefish once inhabited many river systems in great numbers throughout the southern and midwestern states and Montana, New York, and bordering states. Paddlefish once occurred in healthy numbers in at least twenty-six states.

Paddlefish *Polydon spathula* **Ron Pittard**

Black Coral

Antipatharia spp.
Appendix II

Forming slender, plant-like branches that make it look like underwater flora, Black Coral is actually an animal community that forms a delicate, forest-like coral reef in small patches along the ocean bottom.

Black Coral colonies are usually found at depths of sixty feet or more in restricted areas of the Pacific and Atlantic Oceans, and are highly prized by the jewelry trade.

Commercial overharvest — the coral industry is worth hundreds of millions of dollars — threatens to destroy this species. Black Coral also has played a significant role in the illegal collecting by souvenir hunters. Illegal trade activities causing overharvesting are serious threats posing not only economic extinction, but entire species extinction as well. The species is protected under numerous state and island regulations, as well as CITES.

In addition to CITES monitoring, severely limiting or eliminating harvest of Black Coral to prevent overexploitation and public education concerning the vulnerability of this species will enable Black Coral to survive and hopefully thrive once again.

Because of identification problems among various Antipatharia species, the entire order is included for CITES trade protection.

HABITAT:
Black Coral is found in restricted areas along the Atlantic and Pacific shorelines and in the Caribbean.

Black Coral *Antipatharia spp.* **Ron Pittard**

Barn Owl

Tyfo alba
Appendix III

As pest controllers, Barn Owls are highly desirable birds to have around. In a single week during nesting season, a Barn Owl brood and its parents may kill and consume a hundred or more rats and mice, not to mention many undesirable (to man) insects.

Incredibly keen eyesight and hearing — along with the owl's remarkable ability to swivel its head 270 degrees — provide owls with a razor-sharp awareness of all happenings in their domain. The owl's famous ability to turn its head almost completely around is made possible by specialized neck muscles and fourteen neck vertebrae.

These sensory capabilities, coupled with wings capable of near-silent night flight, large feet and talons, and an ominously sharp curved bill are more than ample tools for successful hunting of its favored prey species: small mammals, birds, lizards, frogs, snakes, fish, insects — even bats and kestrels.

Barn Owls are expanding both in numbers and range, and are considered to be a healthy species.

HABITAT:
Barn Owls can be found throughout North America and breed throughout the United States except in the Rocky Mountains. They are highly adaptable and will hunt and nest in both rural and city habitats.

Barn Owl *Tyfo alba* Eric Renk

Burrowing Owl

Athene cunicularia
Appendix III

This pugnacious little owl is only about ten inches long and weighs less than three quarters of a pound. But the Burrowing Owl's size and seemingly quiet demeanor is deceiving. When intruders attempt to enter its desert floor burrow, the Burrowing Owl transforms into a ball of feathered fury, lunging, snapping its bill, screeching, stomping, and even making a rattling sound that closely mimics an aroused rattlesnake.

As with most owls, the Burrowing Owl builds no nest of its own. It takes over the abandoned tunnels of ground squirrels or other burrowing mammals.

Primary food items of the Burrowing Owl are small rodents, insects, reptiles, and small birds which the owl's young ravenously consume inside their air-conditioned burrow.

Loss of desert and grassland habitats required by this owl for nest sites is the primary threat to its survival. There is no trade threat, but some young are taken from the wild for pets, which usually ends in tragedy for the owlet.

The Burrowing Owl is protected by U.S. fish and wildlife laws, Canadian and Mexican laws, and by CITES. All owls are targeted by CITES for protection, even if they are not currently endangered.

HABITAT:
The Burrowing Owl prefers desert or prairie-type landscapes of the western and midwestern states, extending into British Columbia and as far south as Mexico.

Burrowing Owl Athene cunicularia **Eric Renk**

West Indian Manatee

Trichechus manatus
Appendix I

Sometimes referred to as Sea Cows, the Manatees were thought by seamen — probably too long at sea — to be the legendary ocean siren, the mermaid.

The main threat to Manatees in United States waters is boat traffic, which is ever increasing in Florida. Reserves where boating is prohibited and reduced speed limits in known Manatee waters have helped, but Manatee injuries from boat props continue to be documented.

Loss of suitable habitat and pollution also pose serious threats to this species, which is also known as the Florida Manatee.

Outside the United States the Manatee has been hunted throughout Central and South America for its meat, oil, and leather, primarily on a subsistence level.

While there is little trade in Manatees, they face a host of more serious threats, which, combined with low reproductive rates, make their future uncertain.

Manatees are protected by CITES Appendix I, the Marine Mammal Protection Act, the Endangered Species Act and various state laws.

Habitat protection is critical to the survival of Manatees, as is continued monitoring by CITES to ensure against potential trade threats. Public awareness of Manatee locations to avoid injuring the animals with boat props continues to be important in U.S. waters.

HABITAT:
Manatees are found in northern Florida, seasonally in Georgia, and rarely in South and North Carolina. Their range extends through Central America to Brazil and the east coast of Mexico. Manatees are uncommon or rare over most of their range.

West Indian Manatee *Trichechus manatus* **Rusty Rust**

Florida Puma

Felis concolor coryi
Appendix I

Also known as Florida Panther, or Florida Mountain Lion, the Florida Puma is among the most highly endangered mammals found in North America.

The Florida Puma is considered extremely rare, with population estimates as low as several dozen individuals. Reliable population counts are difficult to achieve for this highly elusive and shy cat.

Because of its extremely low numbers, inbreeding poses a difficult problem for the Florida Puma, and it is debatable whether this species will have the genetic base to continue its wild existence for long.

Automobiles and loss of habitat are the main threats facing this species. Illegal hunting now poses a serious threat because of its extremely low population.

Trade threats appear to be minimal. Florida Puma are protected from trade under Appendix I of CITES, and, additionally protected under the Endangered Species Act and state laws.

The Eastern Cougar, a similar looking subspecies, is also listed in Appendix I.

Habitat protection of the puma preserves is essential along with strict enforcement of puma protection laws to maintain this animal's future. The Florida Panther is one of three subspecies of American Mountain Lion.

HABITAT:

This species once existed throughout the south, mainly Florida, Louisiana, Mississippi, Georgia, and Alabama. Today, the only known population is now found in the interior of Florida.

Florida Puma *Felis concolor coryi* Rusty Rust

American Crocodile

Crocodiles acutus
Appendix I

Perhaps most damaging to the future of the American Crocodile is its undeserved reputation as a man-eater, a case of guilt by relation to the larger and more hostile African and Australian Crocodiles.

The population status of this species is unknown, but breeding females are estimated at below sixty, and evidence of young crocodiles and successful nesting is scarce. Despite legal protection and enforcement activities by state, federal and international conservation agencies, the American Crocodile has not responded with favorable population increases.

The American Crocodile was placed on the U.S. List of Endangered Species in September 1975, and listed as endangered throughout its entire range by the IUCN, the World Conservation Union. Under the CITES Appendix I listing trade in American Crocodiles or their parts is illegal. Additionally, the species is protected under Florida state law.

Other countries offer legal protection, but enforcement in most cases is nil. Cuba has set aside a protected crocodile reserve.

The American Crocodile resembles its no-longer-endangered cousin, the American Alligator. The listing by CITES of both species aids in their protection.

If the American Crocodile is to survive as a viable species, it will require a public relations program aimed at awareness and appreciation of crocodiles and their ecological role, continued protection of habitat, intensive study and ongoing monitoring by all agencies concerned.

The American Crocodile once could be found in the United States from south Florida west to Mobile Bay, Alabama. Since adults will wander into saltwater, its range may have included estuary marsh areas outside of Florida.

HABITAT:
Today this highly endangered species is restricted to Everglades National Park, National Florida Key Deer Refuge, and several cooling canals for power generation stations. Outside the U.S., American Crocodile populations are found through Central America to mid-South America, the Mexican coasts, Jamaica, Puerto Rico and Cuba.

American Crocodile *Crocodiles acutus* Rusty Rust

American Alligator

Alligator mississippiensis
Appendix II

The alligator has the distinction of being one of the few reptiles that not only cares for its nest site, but looks after its young after hatching. No reptile species other than alligators and crocodiles are known to care for their young.

Similar in appearance to the highly endangered American Crocodile, the American Alligator has made a remarkable comeback over the last two decades, from the brink of extinction to populations now totaling several million animals. Limited state hunting and trapping seasons on alligators are now in effect, and both domestic and wild hide and meat legally enter U.S. and foreign markets.

The American Alligator is an example of the effect CITES listing can have; the original Appendix I listing is credited by state fish and wildlife officials as a major factor in the decline of illegal hunting and trapping for alligator hides and meat.

The similar appearance between the American Alligator and the endangered American Crocodile is justification for the species' Appendix II listing. Alligators are well managed on a state and federal basis and continued monitoring by CITES to limit trade will keep the alligator in good health as a species.

Under current management practices, including habitat protection, the American Alligator should continue to be a valuable niche in its ecosystem as well as a commercially valuable resource.

Alligators benefit from additional protection under state fish and wildlife laws and through the United States Endangered Species Act.

HABITAT:
The American Alligator once could be found from the swamps of New Jersey through most of the lowlands of the southeastern U.S. and as far west as central Texas. The species' current range is now limited to the southern United States.

American Alligator Alligator mississippiensis **Rusty Rust**

Green Sea Turtle

Chelonia mydas
Appendix I

Graceful in the water, the Green Sea Turtle's dry land abilities leave it vulnerable to poachers, particularly when it comes to egg laying. The turtle lumbers up a sandy beach to the high water mark, scoops out a nest and deposits twelve to several dozen oval-shaped eggs. These nests are an easy mark for poachers, and are one of the reasons the Green Sea Turtle has been targeted for protection.

Historically, Green Sea Turtle numbers were enormous, with an estimated fifty million occurring in the Caribbean alone. World trade in turtle shell and meat, as well as egg collecting, has reduced those wild populations to remnants.

The Green Sea Turtle is protected from trade under Appendix I of CITES, U.S. laws, and various regulations and laws throughout Mexico and Central America.

Sea turtle products such as eggs, hides, oil, and meat are nearly impossible to differentiate. This presents potential protection problems. Maximum trade protection can be given only if all species of sea turtles continue to be listed on Appendix I.

Habitat protection — complete protection of nest sites and eggs in particular — is critical to this species' survival. A major threat to the species comes from a rapidly growing population of raccoons. Sea turtle ranching may prove effective in restoring Green Sea Turtle populations to healthy levels if commercial harvesting is carefully controlled and monitored.

HABITAT:
Before exploitation, the Green Sea Turtle had a worldwide saltwater distribution. Today the species is found in the Gulf of Mexico and the central Gulf of California.

Green Sea Turtle *Chelonia mydas* Rusty Rust

California Condor

Gymnogyps californianus
Appendix I

The condor, a member of the vulture family that can weigh up to twenty pounds, is the largest bird capable of flight. Once an impressive sight as it prowled the skies in search for carrion, the species is on the brink of extinction despite heroic efforts to save it.

Numerous factors have contributed to the condor's extinction in the wild; chiefly DDT and lead poisoning, shooting, trapping, power line accidents and collecting. With no natural predators, the condor's demise is directly attributable to man and the condor's inability to adapt to man-caused changes in its habitat. The condor's recovery is totally dependent upon man's efforts to successfully breed, incubate, and rear captive condor adults and chicks to a point where they may be reintroduced into the wild.

After the last wild condor was captured in 1987, the total captive population stood at twenty-seven birds. Captive bred chicks have since been hatched, and are being released back into the wild.

The goal of the condor captive breeding program is to manage the population to optimize reproductive potential and genetic diversity, for the purpose of reintroducing chicks into the wild. The condor program has had its share of critics, but the status of the species is improving at least in captivity, and hopes run high that successful reintroduction will occur, causing condor to once again soar the California thermals.

The California Condor's continued survival then will depend on having sufficient habitat to meet its needs and protect it from accidental and intentional harm. This is a growing problem as society makes more inroads into the rugged mountains of southern California, where the condors and man are united in a valiant struggle to regain viability for the species.

HABITAT:

The California Condor once ranged over most of western North America, from Canada through Baja California.

California Condor Gymnogyps californianus **John Ruthven**

Imperial Woodpecker

Campeophilus imperialis
Appendix I

This largest of all woodpecker species, noted for its recurved head crest, was found in the pine forests of the Sierra Madre mountains of Mexico. No reliable sightings have been reported since 1956 and expeditions to find the species since then have failed.

Imperial Woodpeckers, frequently called Mexican Ivory Billed Woodpeckers, are presumed extinct. While habitat loss due to logging has played a part in this woodpecker's decline, hunting has been the primary cause of its demise. The species was killed for food and for its bill and feathers, which were thought by natives to have curative health benefits.

Suitable habitat in the form of large pine forest areas still exists in the Imperial Woodpecker's former habitat. However, it is doubtful that the species could ever recover (even if a few breeding pairs still remain) due to the near impossibility of enforcing wildlife laws in such remote areas. If the Imperial Woodpecker still exists, its survival is dependent upon educating and convincing the native inhabitants that the bird must be protected from hunting. Imperial Woodpeckers are protected by Mexican law and, since 1973, CITES Appendix I.

HABITAT:
The Imperial Woodpecker was last sighted in the pine forests of the Sierra Madre mountains of Mexico.

Imperial Woodpecker *Campeophilus imperialis* John Ruthven

Attwater's Prairie Chicken

Tympanuchus cupido attwateri
Appendix I

The Coastal Plains grasses once teemed with probably a million or more Attwater's Prairie Chickens before hunting, habitat loss, changing land uses and practices began taking their toll.

Despite habitat acquisition projects and protective refuges, the Attwater's Prairie Chicken has continued to decline over much of its range. Now restricted to a small area along the Texas coast, and a few islands off the Texas coast, the species is down to less than one thousand birds.

It is not known why recovery efforts have failed to turn this species' precipitous decline around. Evidence suggests, however, that urbanization pressures are a contributing factor. For Attwater's Prairie Chicken to survive, continued acquisition of grassland habitat and refuges will be necessary, along with restocking efforts into suitable habitat. A few birds may be poached from time to time, but otherwise they are not exploited by the commercial trade, nor is sport hunting for the species allowed.

Attwater's Prairie Chicken is protected under state law, the U.S. Endangered Species Act, and CITES Appendix I.

HABITAT:
A thirty-mile-wide strip of land along the coast of southern Texas extending into the middle of Louisiana was the historic belt inhabited by this grouse species.

Attwater's Prairie Chicken *Tympanuchus cupido attwateri* **John Ruthven**

Thick-Billed Parrot

Rhynchopsitta pachyrhyncha
Appendix I

This beautifully marked bird is the only parrot species living in the United States. A small to medium-sized parrot, it was eliminated from Arizona and New Mexico by the early 1900s by a combination of the destruction of its coniferous forest habitat and hunting.

The species' stronghold now is in the Sierra Madre of Mexico. Parrots confiscated by Arizona Game and Fish Department from black market traders have been successfully released into the Coronado National Forest of southeastern Arizona. About a dozen birds currently reside there, and more releases are planned if healthy birds can be obtained.

To this end, a number of zoos and private parrot breeders are captive breeding Thick-Billed Parrots for eventual reintroduction into the wild.

The species is listed under the U.S. Endangered Species Act, state laws, and protected from trade by CITES. Captive breeding programs aimed at restocking this species into suitable habitat, along with protection of migration routes and destinations, offer the most effective means of maintaining wild Thick-Billed Parrot populations.

HABITAT:
Thick-Billed Parrots once ranged the coniferous forests of Mexico with seasonal migrations into Arizona and New Mexico.

Thick-Billed Parrot *Rhynchopsitta pachyrhyncha* **John Ruthven**

Woods Turtle

Clemmys insculpta
Appendix I

Illegal trade and habitat loss have combined to make survival tenuous for the Woods Turtle, a denizen of woodlands and flood plains whose food preferences range from vegetables and berries to insects, worms and scavenged carrion.

Woods Turtles are in considerable demand by collectors, and most of the traded specimens are wild. Very few captive bred Woods Turtles enter the pet trade market.

Despite many state and province laws, endangered species status, and CITES listing, there is still considerable illegal trade both nationally and internationally in this species. The Woods Turtle has declined substantially — or become extinct — in much of its former habitats.

The species is also threatened by pollution, siltation, loss of habitat, and low recruitment of juveniles into the adult population. A scarcity of suitable nesting sites is another factor, compounded by the fact that the few remaining viable sites are often known by collectors and thereby vulnerable to over-collection.

The biggest threat to Woods Turtles is the pet trade, and concerted efforts must continue to educate the pet buying and trading public as to the endangered and declining status of this species.

Some states have designated the Woods Turtle as endangered, but this does not commit the state to protecting habitat.

HABITAT:
Woodland and flood plain environments from eastern Canada (Nova Scotia and New Brunswick) down the eastern coast to Virginia and westward through the Great Lakes states is the habitat preference and range of this long-lived, highly sought-after reptile.

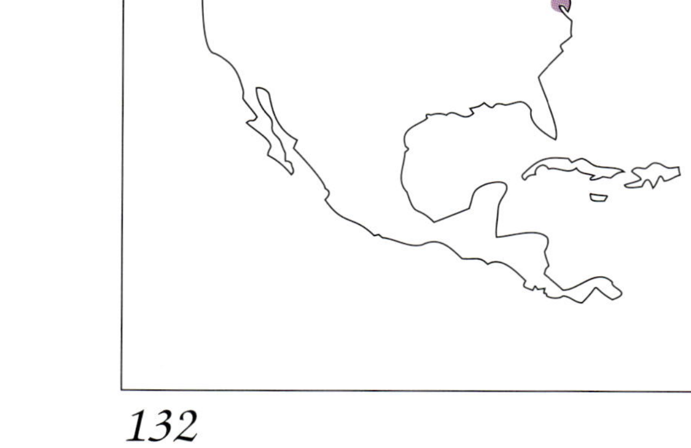

Woods Turtle *Clemmys insculpta* **Patricia Savage**

Gila Monster

Heloderma suspectum
Appendix II

Also known as the Beaded Lizard, the Gila Monster is the only venomous lizard in the world. But this reclusive reptile uses its venom as a defensive protection rather than a weapon of aggression. Lacking fangs, the Gila Monster must literally "chew" its venom into an attacker.

Farming and ranching practices which alter or destroy this lizard's typical mesquite bush habitat are a major factor in its decline. Populations are unknown.

The only known trade threat to the Gila Monster is the pet trade. These exotic-looking lizards are attractively marked, interesting and non-aggressive despite their venomous potential. This makes them desirable to avid reptile collectors and herpetologists.

Trade could become a significant factor if this species becomes in vogue for reptile enthusiasts. Being a reptile, and a poisonous one, at that, does not engender sympathy or much broad public support, which is the plight of most endangered reptiles.

The slow-moving, nonaggressive Gila Monster is protected under state laws and restricted from trade under Appendix II of CITES.

HABITAT:
The Gila Monster is found throughout Mexico to central Arizona. A subspecies is predominant in southeastern Nevada, southwestern Utah, and the Colorado river drainage of western Arizona.

Gila Monster *Heloderma suspectum* Patricia Savage

Red-Kneed Tarantula

Euathlus brachypelma smithii
Appendix II

The Red-Kneed Tarantula, despite its rather fearsome appearance, is docile and rarely bites, even if deliberately provoked. The bite is mildly poisonous, but no more dangerous than a bee sting.

Tarantulas spend most of their time just outside their silk-lined burrows (or, occasionally, silky nests) in scrubland brush, waiting motionless to ambush any hapless prey, such as millipedes, beetles, or grasshoppers which unwittingly venture too close.

The population numbers of this species are not known, and are difficult to ascertain due to low densities and the isolated scrub terrain they frequent. It is thought, however, that areas of land along the western coast of Mexico are probably well populated with Red-Kneed Tarantulas and should remain so in the future, unless development destroys the habitat.

The pet trade could pose a real threat to the species, as there is a growing interest in and demand for tarantulas as exotic pets. Captive breeding programs promise relief to wild-caught specimens, while at the same time satisfying the demands of the exotic pet trade.

This situation bears close monitoring as conservation of wildlife is a relatively low priority in Mexico. Permits are required to collect wildlife in Mexico, but enforcement is difficult. In Mexican livestock markets, with enough money, nearly any Mexican species can be purchased and traded through black markets. A large hairy spider, such as the Red-Kneed Tarantula, could certainly be wiped out from the wild without close scrutiny.

HABITAT:
The Red-Kneed Tarantula is found only in Mexico, (primarily west of the Sierra Madre Occidental).

Red-kneed Tarantula Euathlus brachypelma smithii **Patricia Savage**

Alabama Canebrake Pitcher Plant

Sarracenia rubra spp.alabamensis
Appendix I

The color and scent of the pitcher plant are attractive to insects, but it's a fatal attraction for insects unlucky enough to succumb to it. The plants derive most of their nutritional requirements from their prey, which slides down the slippery interior walls of the plant into digestive liquids.

The Canebrake Pitcher Plant exists in only twelve sites, all of which are privately owned, and probably fewer than five hundred of these plants remain in the wild.

A number of threats to the pitcher plant's survival exist, most of them having to do with development and agricultural practices. A decreasing water table, the reduction in the number of bogs and swamps (often drained for agricultural and land development uses), gravel mining and herbicide use are among the most serious threats to the pitcher plant's survival.

Additionally, interest among plant hobbyists for carnivorous plants is high, and wild plant collecting threatens the few remaining plants. The Canebrake Pitcher Plant, a botanical curiosity, is sought after by collectors on a national and international level.

Since all known sites for the species are privately owned, cooperation between landowners and Canebrake Pitcher Plant recovery efforts are essential. Such efforts include habitat protection and thinning of vegetation which shelters the plant from much-needed sunlight.

The Alabama Canebrake Pitcher Plant is protected by the Endangered Species Act and from trade through CITES Appendix I listing.

Obviously, with so few individual plants left in the wild, measures to eliminate its trade are imperative.

HABITAT:
The Alabama Canebrake Pitcher Plant is endemic only to the swamps and bogs along the fall line of the central Alabama counties of Elmore, Chilton, and Autauga.

Alabama Canebrake Pitcher Plant Sarracenia rubra spp.alabamensis **Patricia Savage**

Olive Ridley Sea Turtle

Lepidochelys olivacea
Appendix I

As with all sea turtles, the Olive Ridley's practice of laying its eggs on sandy beaches above the high tide line leaves them vulnerable to collection, and results in a low hatching rate.

The Olive Ridley was once the world's most abundant sea turtle, numbering hundreds of millions worldwide. But commercial exploitation for turtle products has substantially cut the Ridley's range and numbers.

Although the species is protected by CITES Appendix I listing, the Endangered Species Act, and laws of other countries, illegal harvest continues to occur. Enforcement is difficult because of similarities among all sea turtle products.

International control in sea turtle harvesting, elimination of egg taking, specimen collecting, and continued close monitoring of the species are essential. The black market trade in sea turtle products continues to be high; thousands are still slaughtered on a yearly basis, posing a serious threat to the future of Ridley's Sea Turtles.

Continued establishment of protected reserves, tagging surveys, and hatchery programs remain important. Collecting trade information and effective enforcement of protective regulations is essential.

HABITAT:
The Olive Ridley Sea Turtle's range includes the Gulf of California and the east coast from Georgia to South America.

Olive Ridley Sea Turtle *Lepidochelys olivacea* Patricia Savage

Leatherback Sea Turtle

Dermochelys coriacea
Appendix I

Both scavenger and predator, this huge oceanic reptile feeds on a variety of plant and animal life. Although its massive size requires great amounts of food to sustain the Leatherback, it may go many weeks without food if necessary.

The Leatherback is the largest of all sea turtles, reaching several tons or more, and historically the most widely distributed.

Eggs of the Leatherback are highly esteemed, and appreciated as a rare delicacy, making the species vulnerable to black market trade.

In addition to CITES Appendix I protection, the Leatherback is protected under the U.S. Endangered Species Act, and national laws of numerous countries where it is found. Continued protection from exploitation and enforcement of existing laws are needed to keep Leatherbacks from becoming a casualty of illegal black market trade.

Protection of nest sites is also critical, as is the case in all of the sea turtle populations.

HABITAT:
Scattered numbers of Leatherbacks are found around the Hawaiian Islands and both coasts of Canada. They are found rarely along the east and west coasts of the United States.

Leatherback Sea Turtle *Dermochelys coriacea* Patricia Savage

Brown Bear

Ursus arctos
Appendix II

This largest of all the earth's land predators, also known as the Grizzly Bear, is extinct over most of its range. A few animals still exist in Montana, Idaho, and Wyoming. Canada has a fairly stable Brown Bear population, as does Alaska.

The species was eliminated through much of its former range by hunting, livestock control programs and habitat shrinkage.

Today, Brown Bear are found only in remote wilderness areas, and habitat destruction remains the greatest threat to their existence. Grizzlies occasionally wander onto private ranches, where they may cause problems that end in their elimination.

Rancher/bear conflicts may pose a threat to this species' existence outside of park boundaries in Montana and Wyoming. Hunting is not a threat to Brown Bear populations, as it is tightly controlled and monitored. Alaska and Canada allow sport hunting for this species.

Trade includes hides, claws, and bear bladders illegally smuggled into overseas markets. Conflicts with man and livestock which force the bear's elimination also pose a threat where bear populations are severely depressed.

Brown Bear are protected under the Endangered Species Act, Canadian wildlife laws, and state laws.

Habitat protection, public education and continued monitoring are all-important elements to this bear's survival.

HABITAT:
The Brown Bear once was found throughout Europe and Asia into Alaska, Canada, the western United States, the Great Plains, and as far south as Mexico. The Brown Bear is extinct in Mexico, Oregon, Washington, and California.

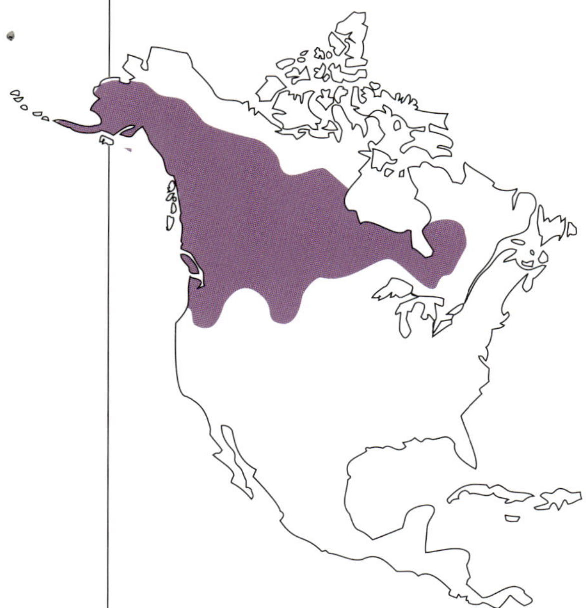

Brown Bear *Ursus arctos* Rita Tate

Black-Footed Ferret

Mustela nigripes
Appendix I

This species has been eliminated over most, and possibly all, of its entire former range. Positive sightings are extremely rare and difficult to verify. Reports of Black-Footed Ferret sightings usually turn out to be other look-alike species.

The extinction of the wild Black-Footed Ferret is a direct result of habitat loss, specifically prairie dog towns, as millions upon millions of prairie dogs were eliminated through pest eradication programs and development of their habitat.

Only a tiny fraction of original prairie dog habitat still remains, and thus the primary food source of ferrets is drastically reduced.

There are no trade threats or potential trade threats to this species.

In addition to CITES Appendix I protection, the Black-Footed Ferret is protected under the Endangered Species Act.

The only method of ensuring the survival of the Black-Footed Ferret is through captive breeding and subsequent reintroduction into suitable habitat.

HABITAT:
The historical range of this scourge of prairie dog towns included all of the Great Plains of the United States and Canada, as far south as Texas, Arizona, and New Mexico.

Black-Footed Ferret *Mustela nigripes* **Rita Tate**

Walrus

Odobenus rosmarus
Appendix III

This toothsome creature is immediately recognized by the two huge white tusks found on both males and females of the species. The Walrus is the only marine mammal to carry these ivory tools, used for feeding — to dig up the ocean bottom in search of clams and other bottom shellfish — and as weapons of defense and dominance.

To complete this animal's "don't mess with me" look is an immense body. Males are around twelve feet long and weigh up to three thousand pounds, and females reach nine feet in length and up to eighteen hundred pounds.

The Walrus traditionally has been of economic importance to the people of the far north, who carved their tusks for ceremonial and commercial purposes and used their meat and hides.

The species is protected under the Marine Mammal Protection Act. The native people of Alaska are allowed a small quota of Walrus.

Under vigorous protection through state, federal, and international laws, the Walrus has made a comeback. Current population is estimated at between fifty and ninety thousand animals.

The main threat to Walrus populations stems from habitat degradation in the form of oil spills and other pollutants. From time to time a few Walrus are still illegally shot on their ice floe resting area. Not all Walrus killed or wounded by native people during subsistence hunts are retrieved, and allocated kills may be higher than actually reported.

HABITAT:
The range of this species includes the ice floes and offshore waters of the Bering Sea, Arctic, and Aleutian Islands.

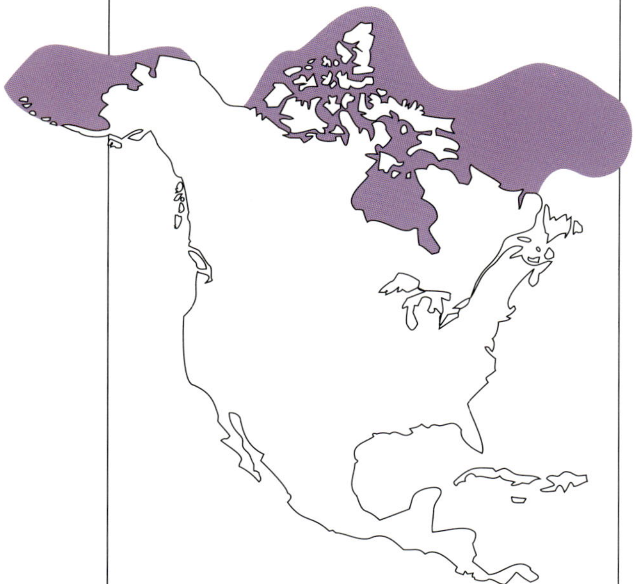

Walrus *Odobenus rosmarus* **Frank Walsh**

Narwhal

Monodon monoceros
Appendix II

It may imply a mythical quality, but the spiraling nine-foot tusk that projects forward from the forehead of the male Narwhal has a more practical purpose in fighting and courtship rituals.

At maturity this toothed whale will reach eighteen feet in length and weigh three thousand pounds. The adult coloration tends toward mottled gray with a white underside. Young of the species are gray and tuskless; they are similar in appearance to young Beluga Whale.

The Narwhal's preferred foods are fish, squid, shrimp, and various crustaceans. It may dive up to twelve hundred feet in search of a meal.

All cetaceans enjoy some measure of protection through international whaling laws, moratoriums and quotas, but Narwhals never were a highly targeted species. They face the same environmental hazards as the Beluga: loss of food sources due to ozone depletion and severe increase in ultraviolet rays in the Arctic regions.

HABITAT:
Narwhal are found only in Arctic and northern sub-arctic seas, in the same waters frequented by Beluga.

Narwhal *Monodon monoceros* **Frank Walsh**

Humpback Whale

Megaptera novaengliae
Appendix I

An operatic virtuoso, the Humpback has gained an appreciative audience for its haunting melodies above the surface of the sea. Musicians have recorded the songs of the whale — which the Humpback uses to communicate information and emotion through the water — and make their own harmonies with the whale's extensive range and melody.

The Humpback has recovered slowly since its harvest was banned in 1966. Population estimates range between nine and twelve thousand. Small numbers are still taken by natives for subsistence purposes.

This species is protected under the International Whaling Commission (IWC), the Marine Mammal Protection Act, the Endangered Species Act, and CITES Appendix I.

Historically, the species was harvested for whalebone, oil and other products, with upwards of fifty to sixty thousand whales taken annually from the 19th Century until 1966.

Habitat protection along migration routes and breeding/calving areas is of critical importance to the health of Humpbacks. Whale watching from boats may need tighter control, especially in breeding areas.

Recent surveys indicate that Humpback numbers are improving, and there may be a stronger population — at least among the Hawaiian Humpback population — than previously thought.

HABITAT:
Various Humpback Whale stocks are found throughout the world between the Arctic and Antarctic Oceans. Generally Humpback populations exhibit seasonal distribution patterns: tropical and subtropical in winter, northern and southern extremes in summer.

Humpback Whale *Megaptera novaeangliae* **Frank Walsh**

Gray Whale

Eschrichtius robustus
Appendix I

A common sight along the Pacific Coast as they migrate close to shorelines — often as close as a few hundred yards from the coast — the Gray Whale has become a new source of revenue to coastal communities hard hit by the collapse of northern fisheries. "Whale watching" charter boat excursions have become a popular tourist attraction in many Pacific Coast communities.

This baleen whale consists of two stocks: the California and the Western or Korean stock.

Taken as far back as Roman times, Gray Whales were hunted to extinction or near extinction over their entire range. The elimination of Gray Whales as whaling targets has seen the species make a strong comeback, from the precipice of extinction to levels approaching historical population highs — about twenty thousand. Due to their unpredictable and sometimes aggressive behavior, this normally peaceful creature was once known as the devilfish.

The Gray Whale's chief concerns are coastal pollution, boat traffic, resource exploration, and oil spills.

While trade in Gray Whales or their derivatives is illegal under numerous national and international laws and moratoriums, they still maintain a high commercial value in revenues paid to whale watching boats, and in nature excursions to calving lagoons.

Protection of Gray Whales is offered through international regulation, the International Whaling Commission (IWC), the Marine Mammal Protection Act, the Endangered Species Act, and CITES. Refuges, especially in Gray Whale calving areas, have proven to be effective conservation measures as increased boat traffic into these areas could cause serious calving problems. The Gray Whale's comeback is secure if all protective measures are enforced and if all forms of coastal pollution and resource exploitation are controlled.

The IWC has classified the Gray Whale as a sustained management stock (California stock), and allows native people a small quota. This poses no threat to the whale's recovery.

HABITAT:
The California stock of Gray Whales migrates along the coastlines of Mexico, the United States, Canada and the former Soviet Union.

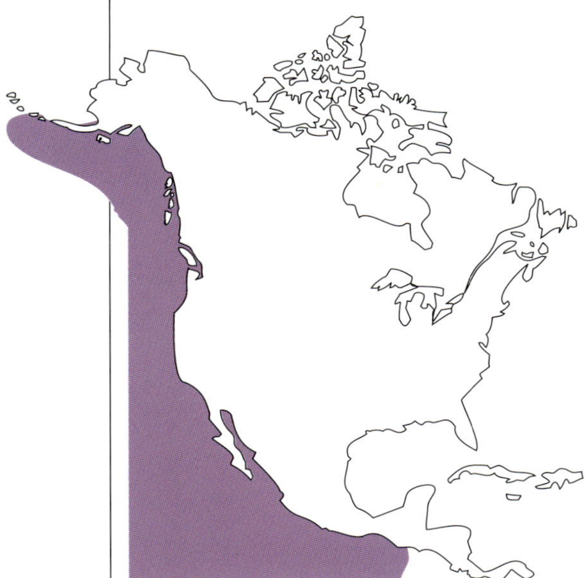

Gray Whale *Eschrichtius robustus* Frank Walsh

Beluga Whale

Delphinapterus leucas
Appendix II

The Beluga, or White Whale, is a medium-sized toothed cetacean up to eighteen feet long and weighing up to four thousand pounds. As the only white whale, and having no dorsal fin, it is one of the more easily identifiable whales. Young Belugas are mottled gray, which turns to bright yellow and, eventually, creme white.

This whale usually travels in pods of a few to several dozen other White Whales during migration. On occasion the pods will foray several miles or more up rivers, for reasons not fully understood.

The diet of the Beluga consists of a variety of fish species, squid, crustaceans and mollusks.

Beluga are protected under international whaling laws and by CITES. However, the species has been in decline due to habitat depredation and reduction of prey items, possibly attributable to the effects of ultraviolet rays, which now penetrate the Arctic region as well as other areas of the world through holes in the ozone layer.

HABITAT:
The Beluga's range is strictly limited to the Arctic and northern sub-arctic seas.

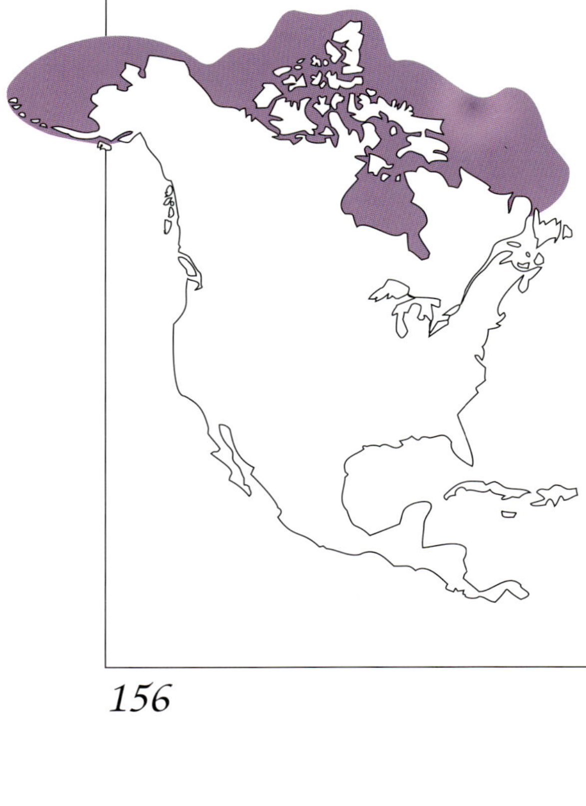

Beluga Whale Delphinapterus leucas **Frank Walsh**

Gulf of California Harbour Porpoise

Phocoena sinus
Appendix I

Comprehensive studies of many porpoise species are lacking, and this is particularly true of the Gulf of California Harbour Porpoise; few researchers have even seen one, and the illustration in this book is one of the few in existence.

More commonly known as Cochita and Vaquita, this porpoise was not recognized as a distinct species until 1958. The population of the Gulf of California Harbour Porpoise is unknown, but presumed to be extremely small, and confirmed sightings are rare.

The Cochitas' main threats are from gill net fisheries, and possibly a food reduction in the upper end of the Gulf of California—this porpoise's only known habitat.

Pollution also poses problems for the survival of Cochita, but the severity of the threat is not known. It is not traded for commercial purposes, nor is there any known commercial utilization for this species. Cochita would be attractive additions to marine theme parks and a unique display, but at this point their live capture is only a potential threat. Cochita are protected by various Mexican laws and by CITES listing Appendix I.

The Cochita's existence is tenuous due to its rarity and limited distribution. Even minor trade would seriously threaten its survival, making the CITES Appendix I listing imperative.

HABITAT:
This marine species is found exclusively in the upper Gulf of California.

Gulf of California Harbour Porpoise Phocoena sinus Frank Walsh

Great Gray Owl

Strix nebulosa
Appendix III

With a wingspan of up to sixty inches and standing just a little shy of three feet tall, the Great Gray is the largest of North America's owl species. However, outward appearances can be deceiving. This raptor's massive feathering conceals a relatively light body of only two or three pounds. The thick coat of feathers is essential for the bird's survival in the tremendously cold temperatures of its habitat.

The Great Gray is a hunter of the dawn and early evening hours, but is more apt to be seen hunting during daylight hours than at night. Primary prey includes mice, voles, squirrels, chipmunks, and other mountain rodents, as well as some birds and reptiles.

When prey is scarce during winter months, Great Grays will migrate into lower elevations in search of food. They may kill and eat small raptors as hunting times and territory overlap.

Owls have a distinctive hunting advantage in that their primary flight feathers have a soft, fringed edge which allows air to pass through noiselessly. The result is deadly, silent, surprise attacks.

Shooting, trapping, pesticides, collisions with power lines and cars all take their toll on this species, but habitat alteration and destruction are the chief threats to its survival. Numbers in the wild are not known, but the Great Gray Owl is believed to be declining. Commercial trade is nonexistent, but feathers are collected for ornamental and ceremonial purposes.

With the exception of the barn owl, most owl species are facing serious problems from the loss of suitable trees for nesting, rearing of young, and hunting.

All owls are protected under CITES as well as by United States fish and wildlife laws.

HABITAT:
The Great Gray Owl's range is restricted to areas of the Cascade and Sierra Nevada mountain ranges in Washington, Oregon, and California, and through the taiga region of British Columbia.

Great Gray Owl *Strix nebulosa* **Sarah Woods**

Whooping Crane

Grus americana
Appendix I

Also known as Whoopers, this tallest of all North American birds was brought perilously close to the brink of extinction by unregulated hunting, loss of breeding range brought on by wetland draining, and collisions with power lines. The species has made a modest recovery, and now numbers three hundred wild and captive birds.

Captive breeding Whoopers are being kept at several zoos, and a few birds have been released in Idaho and at several sites in Florida.

Tremendous efforts to recover the Whooping Crane have been undertaken by the U.S. Fish and Wildlife Service, the Canadian Wildlife Service, the National Aududon Society and through captive breeding programs. Whoopers are protected by the Endangered Species Act, Migratory Bird Act, CITES, Canadian and state laws.

The Whooping Crane's comeback took tremendous effort and coordination between dozens of agencies and institutions, and included a massive public relations program. It is this sort of concerted conservation effort which gives hope to other species in peril.

HABITAT:
The Whooping Crane was once found over most of the continent as far north as the Arctic, south to central Mexico. Today nesting Whooping Cranes occur only in the Wood Buffalo National Park in Canada. Wintering birds are found along the southern coast of Texas, a small area of New Mexico, and Chihuahua, Mexico.

Whooping Crane Grus americana **Sarah Woods**

Woods Bison

Bison athabascae
Appendix I

Shaggier, larger and sporting more massive horns than other bison subspecies, the Woods Bison, or Buffalo, once roamed the central North American plains by the millions — some estimate that more than sixty million once existed.

The species was brought to the brink of extinction by the turn of the century through deliberate and wanton slaughter, partly as an effort to destroy a food source for Native Americans. The Woods Bison began a modest comeback after protection efforts began in 1893. Hybridization and disease have limited this species' numbers to approximately two hundred, all located in the Wood Buffalo National Park in Alberta, Canada.

The Woods Bison's most serious threat comes from diseases carried by other Buffalo species, such as anthrax, which killed many animals in the 1960s.

Protected by the United States government since 1893, Woods Bison are listed on the United States Endangered Species Act and CITES Appendix I and Canadian wildlife laws.

To keep Woods Bison populations viable, isolation from other buffalo subspecies is critical. No other serious threats persist. Other subspecies of United States bison are not endangered or threatened and their populations remain healthy after a remarkable comeback.

HABITAT:
All existing members of this species are located in the Wood Buffalo National Park in Alberta, Canada.

Woods Bison Bison athabascae **Sarah Woods**

Pronghorn Antelope

Antilocapra americana

Appendix I (U.S. populations removed from Appendices, Mexican populations retained in Appendix I).

Historically, the Pronghorn ranged the grassland and prairies of southwestern Canada, United States, and Mexico. It is estimated that before the arrival of the white man, thirty-five million Pronghorns inhabited the continent, but by 1924 over-hunting and other man-caused mortalities had reduced the population to less than twenty thousand.

Hunting restrictions have allowed the United States population to rebound; the species now numbers a half a million animals there, but the Mexican population numbers eight hundred or fewer.

In the United States the species is protected under state fish and wildlife laws, which allow harvesting by strict quota under a permit system. Mexican populations are protected under Mexican law, the United States Endangered Species Act, and CITES.

There is no national trade but a few Pronghorn are legally traded internationally each year as sport trophies. Their only threat is poaching, but effective game law enforcement limits this activity. With the exception of the Mexican populations, the Pronghorn recovery is a wildlife success story unmatched by any other North American mammal.

As long as effective wildlife management techniques continue to be applied along with habitat protection, the Pronghorn should always roam the western plains and grasslands. The Mexican populations will require much more extensive effective law enforcement activities and habitat protection for recovery to be achieved.

HABITAT:
Pronghorn can be found on the grassland and prairies of southwestern Canada, United States, and Mexico.

Pronghorn Antelope *Antilocapra americana* Sarah Woods

Mississippi Sandhill Crane

Grus canadensis pulla
Appendix I

Standing up to four feet tall and with a wingspan of seven feet, the Sandhill Crane is among the largest of North American wading birds. Its darker plumage distinguishes the Mississippi Sandhill Crane from other sandhill crane subspecies.

Feeding upon a variety of plant and animal life including mice, snakes, frogs, and grains, this crane prefers to hunt in fields close to pine or hardwood forests.

Probably never abundant, the species' habitat has been reduced by logging of old growth forests, draining of wetlands and other alterations.

Because the wild Mississippi Sandhill Crane population had been unable to increase, a captive breeding program began in 1966. Birds have been successfully incubated, reared, and released back into the wild, bolstering the wild population to about sixty.

The species is protected on the Mississippi Sandhill Crane National Wildlife Refuge, which was established largely through efforts of the Nature Conservancy. It is listed as an endangered species by the U.S. Fish and Wildlife Service and is listed by CITES for Appendix I.

The continued success of the Sandhill Crane captive breeding program is crucial to the survival of this subspecies.

HABITAT:
The species' original range included the coastal marshes and old growth forests of the Gulf coasts of Florida, Alabama, Mississippi, and Louisiana, but it now consists of southern Jackson County, Mississippi.

Mississippi Sandhill Crane *Grus canadensis pulla* **Sarah Woods**

Credits

The wildlife artists participating in this CITES fund-raising project are all award-winning professionals noted for their unique individual styles and mastery of scientifically exact wildlife art illustration.

A wide-range of artistic styles are represented in this book, from near-photo realism to softer, more emotional, even spiritual approaches. But within this array of styles is one common bond — the illustrations are all scientifically exact renditions of wildlife art.

The list of credits and prestigious awards the artists herein have amassed would fill numerous pages; many of the illustrations included in this book have already been accorded important awards and distinctions. The artwork in this book speaks for itself as one of the finest collections of wildlife art ever published.

E.A. Lusch acted as art director, author and researcher for this book. His company, Windsor Publications, Inc., specializes in biologically exact wildlife charts and prints. Mr. Lusch has been a natural history outdoor writer for 20 years, has authored five books and written hundreds of magazine articles and newspaper columns. He is currently working on several books, including volume two of the CITES endangered species coffee table books. He resides in Eugene, Oregon, with his wife, Linda.

Mark Flint was the editor for the book. He is a media consultant specializing in editing, writing and desktop production for clients in industry, retail, and education. Mr. Flint is currently researching and writing a biography of Glenn Jackson, a leading citizen and dynamic force in Oregon from the 1940s through the 1970s. He has worked as an award-winning reporter and editorial writer for a daily newspaper and also as the publisher of a weekly newspaper. He lives with his wife, Chloe Becca, in Molalla, Oregon.

Carl Darnell designed this publication. He is a graphic designer with regional and national clients. His work encompasses a variety of media, including art direction of a hit PBS series, outdoor murals, computer illustration, logo development, and publication design. In addition, he and his wife, Cindy, operate a mail order company specializing in products and screen printed apparel with themes that promote positive change and diversity. They live and work in the woods of southern Oregon just outside of Shady Cove.

Dr. Ralph Wehinger has been labeled a "visionary" and a "doer" in the Oregon media. He lives with his wife, Laura and their youngest child, Nathan, on their tree farm in Lake Creek, a rural community approximately 25 miles northeast of Medford in the foothills of the Cascade Mountains. He has been a chiropractic physician in private practice in the rural community of Eagle Point since 1977.

Dr. Wehinger's initiative, vision and leadership have brought numerous enhancements to the area. He was instrumental in efforts to rebuild and preserve three covered bridges, and he was a founder of the Lake Creek Historical Society and the International Wildlife Recovery Center.

Dr. Wehinger led the development of the Mark O. Hatfield Environmental Sciences Center in Ashland, Oregon. He also led the effort to develop improvements for the Medford-Jackson County International Airport, including new radar systems, the development of Airport Commerce Park and the establishment of the Jackson County Foreign Trade Zone.

Photo Index of Art & Artists

Arranged alphabetically by common animal name.

Alabama Canebrake Pitcher Plant, 139
Patricia Savage

Aleutian Canada Goose, 79
Jan Martin McGuire

American Alligator, 121
Rusty Rust

American Bobcat, 21
Marian Anderson

American Crocodile, 119
Rusty Rust

American Flamingo, 95
Lawrence B. McQueen

American Ginseng, 63
Norma Hodge

American Kestrel, 57
Laura Mark-Finberg

Atlantic Sturgeon, 105
Ron Pittard

Attwater's Prairie Chicken, 129
John Ruthven

Audubon's Crested Caracara, 37
Guy Crittenden

Bald Eagle, 53
Laura Mark-Finberg

Barn Owl, 111
Eric Renk

Beluga Whale, 157
Frank Walsh

Black Bear, 73
Laney

Black Coral, 109
Ron Pittard

Black-Footed Ferret, 147
Rita Tate

Brown Bear, 145
Rita Tate

Burrowing Owl, 113
Eric Renk

California Condor, 125
John Ruthven

California Pitcher Plant, 71
Norma Hodge

Calypso Orchid, 69
Norma Hodge

Canadian Lynx, 23
Marian Anderson

Cooper's Hawk, 31
Guy Crittenden

Eastern Cougar, 17
Al Agnew

Eskimo Curlew, 93
Lawrence B. McQueen

Florida Puma, 117
Rusty Rust

Gila Monster, 135
Patricia Savage

Golden Eagle, 87
Jan Martin McGuire

Gopher Tortoise, 35
Guy Crittenden

Gray Whale, 155
Frank Walsh

Gray Wolf, 15
Al Agnew

Great Gray Owl, 161
Sarah Woods

Great Horned Owl, 41
Wayne Dowdy

Green Sea Turtle, 123
Rusty Rust

Gulf of California Harbour Porpoise, 159
Frank Walsh

Gyrfalcon, 83
Jan Martin McGuire

Harpy Eagle, 61
Laura Mark-Finberg

Hawaiian Hawk, 89
Lawrence B. McQueen

Humpback Whale, 153
Frank Walsh

Imperial Woodpecker, 127
John Ruthven

Jaguar, 11
Al Agnew

Jaguarundi, 51
Laura Mark-Finberg

Laysan Duck, 91
Lawrence B. McQueen

Leatherback Sea Turtle, 143
Patricia Savage

Margay, 29
Joey Berlin

Masked Bobwhite, 39
Guy Crittenden

Mexican Bighorn Sheep, 47
Wayne Dowdy

Mexican Bobcat, 43
Wayne Dowdy

Mississippi Sandhill Crane, 169
Sarah Woods

Montezuma or Harlequin Quail, 97
Lawrence B. McQueen

Narwhal, 151
Frank Walsh

Northern Goshawk, 59
Laura Mark-Finberg

Northern Harrier, 45
Wayne Dowdy

Ocelot, 27
Joey Berlin

Olive Ridley Sea Turtle, 141
Patricia Savage

Onconee Bells, 67
Norma Hodge

Osprey, 75
Laney

Paddlefish, 107
Ron Pittard

Peregrin Falcon, 55
Laura Mark-Finberg

Polar Bear, 81
Jan Martin McGuire

Prairie Falcon, 13
Al Agnew

Pronghorn Antelope, 167
Sarah Woods

Queen Conch, 101
Ron Pittard

Red-Kneed Tarantula, 137
Patricia Savage

Red-Tailed Hawk, 77
Jan Martin McGuire

River Otter, 85
Jan Martin McGuire

Saw Whet Owl, 25
Joey Berlin

Scarlet Ibis, 49
Wayne Dowdy

Short-Nosed Sturgeon, 103
Ron Pittard

Short-Tailed Albatross, 99
Lawrence B. McQueen

Siskiyou Lewisia, 65
Norma Hodge

Snowy Owl, 19
Marian Anderson

Southern Sea Otter, 33
Guy Crittenden

Thick-Billed Parrot, 131
John Ruthven

Walrus, 149
Frank Walsh

West Indian Manatee, 115
Rusty Rust

Whooping Crane, 163
Sarah Woods

Woods Bison, 165
Sarah Woods

Woods Turtle, 133
Patricia Savage